Morphofunctional Aspects of Tumor Microcirculation

Domenico Ribatti

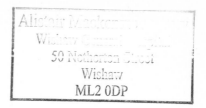

Morphofunctional Aspects of Tumor Microcirculation

 Springer

Prof. Dr. Domenico Ribatti
Policlinico, Department of Basic Medical Sciences
Section of Human Anatomy and Histology
University of Bari Medical School
Bari
Italy

ISBN 978-94-007-4935-1   ISBN 978-94-007-4936-8 (eBook)
DOI 10.1007/978-94-007-4936-8
Springer Dordrecht Heidelberg London New York

Library of Congress Control Number: 2012949567

Printed on acid-free paper

Springer is part of Springer Science+Business Media (www.springer.com)

Acknowledgments

The research leading to these results has received funding from the European Union Seventh Framework Programme (FP7/2007–2013) under grant agreement n°278570

Contents

List of abbreviations

AE	Adverse effects
Ang	Angiopoietin
AP	Aminopeptidase
APC	Antigen presenting cell
ATP	Adenosine triphosphate
BLI	Bioluminescence imaging
CAM	Chorioallantoic membrane
CI	Confidence interval
COX	Cycloxygenase
CSF	Colony stimulating factor
DC	Dendritc cell
DXR	Doxorubicin
EDGF-1	Endothelial differentiation gene-1
EGFR	Epidermal growth factor receptor
eNOS	Endothelial nitric oxide synthase
EPC	Endothelial precursor cell
EPH	Ephrin
FDA	Food and Drug Administration
FGF-2	Fibroblast growth factor-2
G-CSF	Granulocyte-colony stimulating factor
GFP	Green fluorescent protein
GIST	Gastrointestinal stromal tumors
GM-CSF	Granulocyte macrophage-colony stimulating factor
GRO	Growth related oncogene
HGF	Hepatocyte growth factor
HIF	Hypoxia inducible factor
HPV	Human papilloma virus
HSC	Hematopoietic stem cells
IFL	Irinotecan 5-fluorouracil and leucovirin
IFN	Interferon
IL	Interleukin
IMG	Intussusceptive microvascular growth

LDL	Low density lipoprotein
MAPC	Multipotent adult progenitor cells
MBP	Major basic protein
MCP	Monocyte chemoattractant protein
MDSC	Myeloid- derived suppressor cells
MHC	Major histocompatibility complex
MIP	Macrophage inflammatory protein
MMP	Matrix metalloproteinase
MRD	Minimal residual disease
MRI	Magnetic resonance imaging
NECA	(N-ethylcarboxámido) adenosine
NGF	Nerve growth factor
NK	Natural killer
NRP	Neuropilin
NSCLC	Non small cell lung cancer
OS	Overall survival
PA	Plasminogen activator
PAR	Proteinase activated receptor
PDGF	Platelet derived growth factor
PDGF-R	Platelet derived growth factor-receptor
PF4	Platelet factor 4
PFS	Progression-free survival
PG	Prostaglandin
PlGF	Placental growth factor
PR	Partial response
PSMA	Prostate-specific membrane domain
PyMT	Polyoma middle-T
RIP1-TAG2	Rat insulin promoter-1 T-antigen-transgene 2
RTK	Receptor tyrosine kinase
SAGE	Serial analysis of gene expression
SCID	Severe combined immunodeficiency
SD	Stable disease
SDF	Stromal-cell derived factor
SKITL	Soluble KIT ligand
SR	Stable response
STAT3	Signal transducer and activator of transcription 3
TAM	Tumor associated macrophage
TDI	Toluene diisocyanate
TEM	Tie-2 expressing monocytes
TGF-β	Transforming growth factor-β
TKI	Tyrosine kinase inhibitors
TIMP	Tissue inhibitor of matrix metalloproteinase
TNF-α	Tumor necrosis factor α
TSP	Thrombospondin
TSP	Thrombospondin repeat

VDA	Vascular disrupting agents
VE-cadherin	Vascular endothelial cadherin
VEGF	Vascular endothelial growth factor
VEGFR	Vascular endothelial growth factor receptor
VTA	Vascular targeting agents
VVO	Vesciculo-vacuolar organelles
vWF	von Willbrand factor
ZO	Zonula occludens

Chapter 1
Introduction

The human vascular system is composed of a network of vessels lined by endothelial cells. In addition to its critical role in gas exchange, blood components are distributed to tissues via this route, including nutrients, hormones, growth factors, inflammatory mediators, and inflammatory cells.

The normal microvessels consist of arterioles, capillaries, and venules, and form a well-organized, regulated and functional architecture and are characterized by dichotomous branching. Arteries and veins of different size share many similarities of architecture and functional organization. Veins are more numerous and generally have larger caliber and thinner walls with less muscle and elasticity that their arterial counterparts, because they carry more volume at lower pressure. Each vessel usually has an intima with luminal endothelial cells characterized by peculiar properties (Table 1.1) and subendothelial fibroelastic connective tissue, a media formed of smooth muscle cells, elastic fibers and collagen, and the adventitia formed of connective tissue, nerves and capillaries (vasa vasorum).

Capillaries are the blood vessels with the smallest diameter, but they have a great surface area. Their basic structure consists of an endothelial cells layer, a basement membrane, and supporting cells termed pericytes. There are three types of capillaries: continuous, fenestrated, and sinusoidal. Fenestrations are approximately circular, 50–100 nm in diameter, and at their edge the luminal and abluminal membranes of the endothelial cells come in contact with each other. Fenestrated capillaries occur in renal glomeruli and in endocrine glands. Sinusoids are expanded capillaries and are large and irregular in shape. They have true discontinuities in their walls, allowing intimate contact between blood and parenchyma. Sinusoids occur in liver, spleen and bone marrow. Continuous capillaries occur in the brain, striated and smooth muscles.

In normal condition, the gross vascular anatomy of the vascular system is characterized by a reproducible branching pattern. Control of branch patterning includes both attractive and repulsive guidance signals (Childs et al. 2002) and is regulated by both positive and negative regulators (Lee et al. 2001).

Vascular architecture in tumor is different from normal tissues (Table 1.2) and is determined by micro heterogeneities in the cellular interactions with the extracellular matrix. In tumors, the organ- and tissue-specific vascular architecture is not retained. The highest vessel densities are usually found in the periphery within the invasive

D. Ribatti, *Morphofunctional Aspects of Tumor Microcirculation*,
DOI 10.1007/978-94-007-4936-8_1, © Springer Science+Business Media Dordrecht 2012

Table 1.1 Properties of endothelial cells

Regulation of blood flow via dilatation and constriction of vessels
Modulation of selective permeability and substance transfer
Gas, nutrient, and waste exchange
Lipid metabolism
Extracellular matrix production and modulation
Growth factors secretion and regulation
Smooth muscle cell regulation
Regulation of inflammatory reactions
Immunologic functions
Coagulant and thrombotic functions

Table 1.2 Morphological and functional characteristics of the vasculature in normal tissue and tumor

Global organization
Normal (normal)
Tumor (abnormal)
Pericyte coverage
Normal (normal)
Tumor (absent or detached)
Basement membrane
Normal (normal)
Tumor (absent or too thick)
Vessel diameter
Normal (normal distribution)
Tumor (dilated)
Vascular density
Normal (normal, homogenous distribution)
Tumor (abnormal, heterogeneous distribution)
Permeability
Normal (normal)
Tumor (high)

front. The architecture of the tumor seems to be primarily determined by the tumor cells themselves.

Tumor blood vessels do not display the recognizable features of arterioles, capillaries or venules, are irregular in size, shape, and branching pattern, form arteriovenous shunts. Their haphazard branching patterns and larger, less regular diameters contribute to the non-uniform perfusion of tumor cells. Moreover, they have uneven diameters, chaotic flow patterns, and increased permeability to macromolecules (McDonald and Baluk 2002). Tumor vessel density is very heterogeneous: the highest values are found in the invading tumor edge, where the density is four to ten times greater than inside the tumor (Giatromanolaki et al. 2002) and the arrangement of vessels in the centre of a tumor is much more chaotic than at its edges.

Tumor vessels exhibit a spectrum of vessel subtypes ranging from capillaries and 'mother' vessels (large, leaky, thin-walled, pericyte-depleted fenestrated sinusoids) to glomeruloid vessel (poorly organized vascular structures that macroscopically resemble renal glomeruli, composed of endothelial cells and pericytes with minimal vascular lumens and reduplicated basement membrane) outgrowth and vascular malformations (mother vessels that have acquired an asymmetric coat of smooth muscle cells and/or fibrous connective tissue) (Nagy et al. 2010).

For example, in healthy liver, sinusoidal endothelial cells are fenestrated, separated by wide junctions and devoid of basement membrane. By contrast in hepatocellular carcinoma, endothelial cells become capillarized, with fewer fenestrations, more junctions, and abundant perivascular matrix; these features reduce perfusion and exchange of oxygen (Yang and Poon 2008). Hepatocellular carcinoma patients with sinusoid-like vasculature have a shorter survival time, although the microvascular density of a tumor with sinusoid-like vasculature is significantly less than that of a tumor with capillary-like microvessels (Chen et al. 2011).

The molecular mechanisms causing abnormal vascular architecture are not completely understood, but the imbalance of pro- and anti-angiogenic factors is considered to be a key contributor. Moreover, mechanical stress generated by proliferating tumor cells also compress vessels in tumors (Padera et al. 2004), with some vessels being oversized, other being more immature smaller vessels (Fukumura et al. 2010).

These structural abnormalities result in disturbed blood flow, hypoxia, hyperpermeability, and elevated interstitial pressure in many solid tumors, responsible, in turn, of impaired delivery of anticancer drugs as well as oxygen to the tumor site.

Chapter 2
Morphological Aspects of Tumor Vasculature

2.1 Tumor Endothelial Cell Features

Endothelial cells of mature, quiescent vessels are characteristically low proliferative and their estimated turnover times are measured in years, whereas those of tumor vessels are markedly dependent on growth factors for survival. Vascular endothelial growth factor (VEGF) has been convincing assigned a central role in the induction of host vessels into a growing tumor (Tables 2.1, 2.2). When endothelial cells invade a newly formed tumor, they come into contact with tumor cells that produce VEGF, which may be responsible not only for vascular proliferation, but also for the altered permeability of the newly formed vessels.

Tumor endothelial cells proliferate 50–200 times faster than normal endothelial cells (Vermeulen et al. 1995). About 10 % of endothelial cells in tumors were labeled by autoradiography using tritiated thymidine or by staining with bromodeoxyuridine, while in normal tissues ony about 0.2 % of the endothelial cells were labeled (Tannock 1968; Denekamp 1982). Tumor endothelial cell proliferation rate in some regions of the tumor vasculature reflects the angiogenesis that accompanies an increase in tumor volume, whereas in other regions tumor endothelial cells undergo apoptosis in parallel with tumor necrosis and vessel regression.

The tumor-associated endothelium is structurally defective (Fig. 2.1). Discontinuities or gaps (<2 μm in diameter) that allow hemorrhage and facilitate permeability are common features. Cells contacts are usually poorly differentiated and no complex contact structures exist and defects in endothelial cell barrier function, due to abnormal cell-cell junctions and other changes, exaggerate leakiness. This correlates with histological grade and malignant potential (Daldrup et al. 1998) and can be exploited in locating tumors by imaging contrast media and in the delivery of macromolecular therapeutics (Mc Donald and Choyke 2003). Furthermore, it results in extravasation of plasma proteins and even erythrocytes and may facilitate the traffic of tumor cells into the bloodstream and the formation of metastases (Dvorak et al. 1988). Leakiness has been attributed to highly active angiogenesis and microvascular remodeling, but its structural basis and mechanism are unclear. Intercellular gaps, transendothelial holes, vesicle-vacuolar organelles (VVO), and endothelial fenestrae are all present in the endothelium of tumor vessels (Dvorak et al. 1988). Overall perfusion rates

D. Ribatti, *Morphofunctional Aspects of Tumor Microcirculation*,
DOI 10.1007/978-94-007-4936-8_2, © Springer Science+Business Media Dordrecht 2012

Table 2.1 Mammalian VEGF ligands and biological functions

Ligands	VEGFR-binding	NRP-binding	Biological functions
VEGF-A 165	VEGFR-1, R-2	NRP-1, -2	Angiogenesis
VEGF-A 121	VEGFR-1, R-2	NRP-1	Angiogenesis/anti-angiogenesis
VEGF-A 145	VEGFR-1, R-2	NRP-2	Angiogenesis
VEGF-A 189	VEGFR-1, R-2	NRP-1	Angiogenesis
VEGF-B	VEGFR-1	NRP-1	Fatty acid uptake in endothelial cells of the heart
VEGF-C	VEGFR-2, R-3	NRP-2	Lymphangiogenesis
VEGF-D	VEGFR-2, R-3	NRP-2	Lymphangiogenesis

Table 2.2 Direct and indirect effects exerted by VEGF within the tumor microenvironment

Endothelial cells
Survival
Proliferation
Migration
Permeability

Tumor cells
Survival
Motility
Invasiveness

Pericytes
Vessel maturation

Dendritic cells
Maturation

T-cells
Immune functions

Endothelial precursor cells
Recruitment

VEGFR-1+ myeloid cells
Recruitment

are lower in many normal tissue and the average red blood cell velocity in tumor vessels can be an order of magnitude lower than in normal vessels. The heterogeneity of tumor blood flow causes abnormal microenvironment in tumors and hinders the delivery and efficacy of therapeutic agents. Blood flow is variable between different tumors, between the primary cancer and its metastatic lesions, and within the same tumor (Fukumura et al. 2010). Normally, fluid pressure inside healthy vessels is higher than in the surrounding milieu. However, owing to the leakiness of tumor vessels, escaping fluid raises the interstitial fluid pressure which may exceed ten times that of normal tissues and produces edema in and around tumor tissues (Baluk et al. 2003). This condition is aggravated by poorly functioning lymphatic vessels, which are compressed by cancer and stromal cells (Padera et al. 2004).

Tumor endothelial cells protrude extensions into the lumen and form abluminal sprouts, with leading tip cell penetrating deep into the tissue. Tumor -derived micro vesicles promote endothelial cell migration, invasion, and *in vivo* neovascularization. Transfer of CD147 extracellular matrix metalloproteinase (MMP) inducer, a

Fig. 2.1 a Ultrastructural
finding in the stroma of
high-grade lymphoma. A new
capillary with a slit-like
lumen (*arrow head*), formed
by two endothelial cells
arranged in parallel.
b Ultrastructural finding in
the stroma of a low-grade
lymphoma. A fenestrated
capillary, showing flat
endothelial cells (EC) with a
fenestra (*arrow head*).
Original magnification:
a ×16.200; **b** ×4,800.
(Reproduced from Ribatti
et al. 1996)

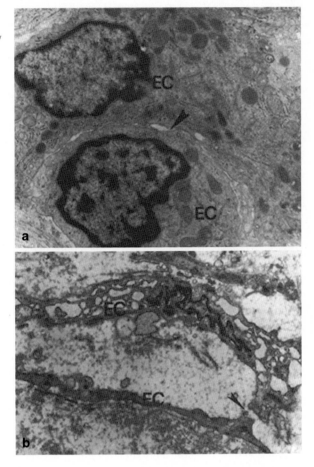

membrane-spanning molecule highly expressed in tumor cells, by micro vesicles
shed by ovarian tumor cells activates endothelial cells and stimulates angiogenesis
(Millimaggi et al. 2007). Moreover, the treatment of ovarian cancer cells with small
interfering RNA against CD147 reduces the angiogenic potential of micro vesicles
(Millimaggi et al. 2007). Al-Nedawi et al. (2008, 2009) suggested that the angiogenic switch in tumors may be induced by micro vesicles-mediated transfer of the
oncogenic epidermal growth factor receptor (EGFR).

Tumor vessels are almost completely covered by basement membrane. In normal
condition, type IV collagen and laminin form two-independent, three-dimensional
network, respectively. Type IV collagen confers structural stability to the basement
membrane, while the different isoforms of laminins are involved in signaling to various tissues. Other minor molecules include collagen types VIII, XV, and XVIII,
thrombospondins (TSP) I and II, and osteonectin. The basement membrane that en-
velops endothelial cells and pericytes of tumor vessels may have extra layers that

have no apparent association with the cells and has an aberrant thickness (Baluk et al. 2003). It may also be focally disrupted with holes and project wide extensions deep in the perivascular interstitium and is composed of other types of extracellular matrix components (Baluk et al. 2003). It contains distinctive forms of fibronectin comprising the ED-B domain, and type IV collagen with exposed cryptic sites (Santamaria et al. 2003), and is a source of angiogenic and anti-angiogenic molecules (Kalluri 2003). Some of its structural proteins are broken down by enzymes to yield molecules with potent actions. Three examples are endostatin, which is a COOH- terminal fragment of collagen XVIII, tumstatin and arresten, which are the noncollageneous-1 domain of the α3 and the α1 chains of type IV collagen respectively (Colorado et al. 2000).

Tumor endothelial cells produce increased levels of proteinases, adhesion molecules and other factors that facilitate diapedesis (Sullivan and Graham 2007). Tumor endothelial cells also differ from those of normal vessels in other ways, including cell surface markers, including aminopeptidase A (APA), aminopeptidase N (CD13), annexin A1, delta-like 4, nucleolin, phospahtidylserine, and the profile and level of cell adhesion molecule they express. They preferentially overexpress the cell-surface molecules integrin $\alpha v\beta 3$ and $\alpha v\beta 1$, E-selectin, endoglin, endosialin (tumor endothelial marker-1) and VEGF receptors (VEGFR), all of which stimulate endothelium adhesion and migration (Magnussen et al. 2005). Tumors treated with antagonists of integrin αv $\beta 3$ display increased apoptosis and vessel regression. One feature of the role of integrin αv $\beta 3$ is that it mediates endothelial cell adhesion to extracellular matrix molecules. Vascular endothelial cadherin (VE-cadherin) is poorly expressed in tumor vessels and it results in their destabilization and may lead to abnormal remodeling.

Tumor endothelial cells present many abnormalities in gene expression. St Croix et al. (2000) carried out an unbiased gene expression analysis (serial analysis of gene expression, SAGE) of endothelial cells immunopurified from human colorectal cancer and human colon and found that tumor-activated endothelium overexpressed specific transcripts as the result of qualitative differences in gene profiling compared with the normal endothelial cells of the tissue of origin. A total of 79 transcripts were differentially expressed: 46 were elevated at least ten-fold and 33 were expressed at substantially lower levels in tumor-associated endothelial cells. Similar expression pattern were found in tumor-associated endothelial cells from metastatic lesions and other primary tumor sites. Most of the differentially expressed genes were also found during luteal angiogenesis and wound healing, which suggests that tumor angiogenesis uses the same signaling pathways as physiologic angiogenic states. These data clearly delineate the effects of the local microenvironment on gene expression patterns in endothelial cells and support the critical role of the microenvironment in defining an angiogenic phenotype.

In another study, SAGE analysis of brain tumor endothelial cells revealed the existence of 14 glioma endothelial markers, up-regulated in endothelial cells isolated from grade III/IV gliomas, 12 of which defined as cell surface or secreted molecules (Madden et al. 2004). SAGE has been also used to compare the transcriptional profile of tumor endothelial cells and endothelial cells from normal liver or regenerating liver (Seaman et al. 2007). Several genes were common to both tumors and regenerating livers and the expression of 13 genes was increased by >10-fold in the tumor samples compared with normal or regenerating liver.

DNA microarray analyses were used to characterize the transcriptome of tumor endothelial cells from ovarian cancer, revealing 400 differentially expressed genes, some of which were validated at the level of protein expression and function (Lu et al. 2007). In breast cancer, tumor endothelial cells expressed specific marker, such as transcription factors SNAIL1 and HEYL, which was restricted to invasive breast tumor endothelial cells, suggesting that the transcriptional profile of tumor endothelial cells is stage-specific (Parker et al. 2004).

Ria et al. (2009) carried out a comparative gene expression profiling of multiple myeloma endothelial cells (MMECs) and monoclonal gammopathies of undetermined significance (MGUS) endothelial cells (MGECs) with Affymetrix U133A Arrays, and demonstrated that 22 genes are differentially expressed (14 down-regulated and 8 up-regulated) at relatively high stringency in MMECs compared with monoclonal gammopathy of undetermined significance endothelial cells (MGECs) (Fig. 2.2). Deregulated genes are mostly involved in extracellular matrix formation and bone remodeling, cell adhesion, chemotaxis, angiogenesis, resistance to apoptosis, and cell-cycle regulation. Validation was focused on *DIRAS3*, *SERPINF1*, *SRPX*, *BNIP3*, *IER3*, and *SEPW1* genes, which were not previously found to be functionally correlated to the angiogenic phenotype of MMECs. The small interfering RNA knockdown of the up-regulated genes *BNIP3*, *IER3*, and *SEPW1* affected critical MMEC functions mediating the cell angiogenic phenotype, that are proliferation, apoptosis, adhesion, and capillary tube formation.

Tumor endothelial cells exhibit chromosomal translocation in B cell lymphomas (Streubel et al. 2004), the BCR/ABL fusion gene in chronic myelogeneous leukemias (Gunsilius et al. 2000) and the myeloma specific 13 q14 chromosomal deletion (Rigolin et al. 2006). In tumors of the nervous system, tumor endothelial cells present the same genetic amplification or chromosomal aberrations of the tumor of origin (Pezzolo et al. 2007; Ricci-Vitiani et al. 2010). In human xenografts of renal carcinoma, melanoma, and liposarcoma, murine tumor endothelial cells are aneuploid, bearing alterations similar to those observed in human tumor endothelial cells (Hida et al. 2004).

Embryonic genes are expressed by the tumor-derived endothelial cells (Bussolati et al. 2003, 2006). Tumor endothelial cells from renal carcinomas express the embryonic renal transcription factor PAX2, which contributes to both apoptosis resistance and pro-angiogenic properties of tumor endothelial cells (Fonsato et al. 2006).

Tumor endothelial cells undergo endothelial to mesenchymal transition and move away from their resident site (Potenta et al. 2008). Epithelial to mesenchymal transition is a characteristic feature of epithelial cell tumor progression. It is characterized by several molecular changes that include the loss of epithelial markers such as E-cadherin and zonula occludens-1 (ZO-1), and the induction of mesenchymal markers such as N-cadherin, fibronectin, vimentin, and Snail.

Fig. 2.2 Unsupervised **(a)**
and **(b)** supervised analysis
of gene expression profiles
from a data set composed of
five monoclonal gammopathy
of undetermined significance
endothelial cells, five multiple
myeloma endothelial cells,
and five HUVEC samples. In
(b) identification of the 22
genes differentially expressed
in five multiple myeloma
endothelial cells versus five
monoclonal gammopathy of
undetermined significance
endothelial cells. Color scale
bar, the relative
gene-expression changes
normalized by the SD; the
color changes in each row is
gene expression relative to
the mean across the samples
(with gene symbols).
(Reproduced form Ria et al.
2009)

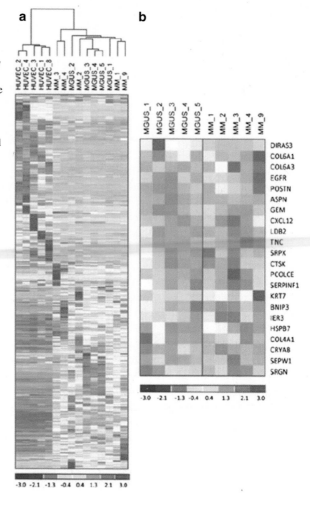

2.2   Tumor Pericytes

Pericytes are perivascular cells that are thought to surround and stabilize new vessels
(Ribatti et al. 2011). They extend thin processes around and along the microvascular
tubes and have areas of direct contact with endothelial cells. Endothelial cells secrete platelet derived growth factor B (PDGF-B), that causes pericyte precursor cell
proliferation and migration through activation of PDGF receptor-β (PDGFR-β). By
contrast, endothelial cells in vascular sprouts release VEGF, which in turn mediates
suppression of PDGFR-β signaling through the induction of VEGFR-2/PDGFR-β
complexes. This pathway abrogates pericyte coverage of endothelial sprouts leading
to vascular instability and regression (Fig. 2.3). Vessel sprouts cause destabilization
of pericyte investment through angiopoietin-2 (Ang-2)/Tie-2 signaling. Pericytes
provide guidance for endothelial movement and tube formation through secretion

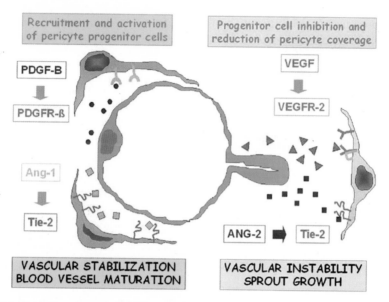

Fig. 2.3 Signaling pathways operating in endothelial cell/pericyte paracrine cross-talk. Vessel sprouts (*right*) cause destabilization of pericyte investment trough Ang-2/Tie-2 signaling. Pericyte provide guidance for endothelial movement and tube formation through secretion of VEGF and soluble NG-2. Spreading endothelial cells, in turn, stimulate pericyte precursor cell proliferation and migration by releasing VEGF and NO. Vessel stabilization (*left*) occurs by pericyte investment and close interaction with endothelial cells. Mature endothelial cells secrete PDGF-B, which promotes proliferation and migration of pericyte precursor cells through activation of PDGFR-β expressed on the surface of pericyte progenitors. This mechanism leads to pericyte coverage of early endothelial tubes. Vessel maturation further develops through Ang-1- and Notch-mediated signaling. Pericyte stabilize and reinforce the endothelial tube contributing to secretion of basement membrane. (Reproduced from Ribatti et al. 2011)

of VEGF and nerve/glial antigen-2 (NG-2). Spreading endothelial cells, in turn, stimulate pericyte precursor cell proliferation and migration by releasing VEGF and nitric oxide (NO) (Fig. 2.3). Vessel stabilization occurs by pericyte investment and close interaction with endothelial cells. Mature endothelial cells secrete PDGF-B, which promotes proliferation and migration of pericyte precursor cells through activation of PDGFR-β expressed on the surface of pericyte progenitors. This mechanism leads to pericyte coverage of early endothelial tubes. Vessel maturation further develops through Ang-1- and Notch-mediated signaling and pericyte stabilize and reinforce the endothelial tube contributing to secretion of basal membrane (Fig. 2.4).

Vessel coverage by pericytes is usually more extensive in tissues with a slow endothelial cell turnover (Diaz-Flores et al. 2009), which explains why tumor vessels often have fewer pericytes. In tumors, pericytes have an abnormal shape and express markers of more immature, less contractile mural cells (Raza et al. 2010). They are revealed by immunohistochemical staining of sections (Morikawa et al. 2002) and may cover 73–92 % of endothelial sprouts in several murine tumor types. Breast and colon tumors recruit significantly more pericytes than gliomas or renal cell carcinoma (Eberhard et al. 2000).

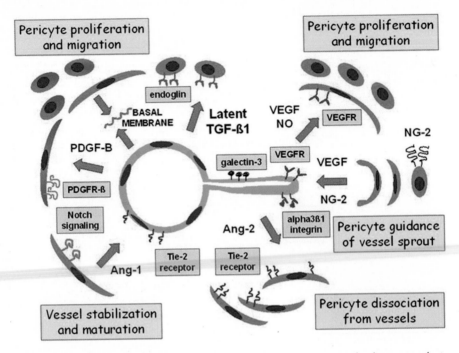

Fig. 2.4 Schematic drawing that illustrates the paracrine interactions occurring between pericyte precursor cells and endothelial cells in PDGF-mediated angiogenesis. Endothelial cells secrete PDGF-B, that causes pericyte precursor cell proliferation and migration through activation of PDGFR-β. Pericytes surround and cover early endothelial tubes. By contrast, endothelial cells in vascular sprouts release VEGF, which in turn mediates suppression of PDGFR-β signaling through the induction of VEGFR-2/PDGFR-β complexes. This pathway abrogates pericyte coverage of endothelial sprouts leading to vascular instability and regression. (Reproduced from Ribatti et al. 2011)

In tumors, pericytes are located near blood vessels at the growing front of tumors, where angiogenesis is most active and show morphological abnormalities (Schlinge-mann et al. 1990; Wesseling et al. 1995; Morikawa et al. 2002). Moreover, pericyte deficiency could be partly responsible for vessel abnormalities in tumor blood vessels (Gerhardt and Semb 2008) and partial dissociation of pericytes (Hobbs et al. 1998; Hashizume et al. 2000) contribute to increased tumor vascular permeability.

Most tumor pericytes are loosely associated with endothelial cells, have abnor-mal shape, paradoxically extend cytoplasm processes away from the vessel wall, and have extra layers of loosely fitting basement membrane (Morikawa et al. 2002). Their abnormalities are consistent with alterations in PDGF signaling pathways. Mice genetically deficient in PDGFB or its receptors have blood vessels with loose pericytes attachment, irregular vessel caliber, luminal projections on endothelial cells, and hemorrhage. Similar abnormalities occur in many tumor vessels. Pericytes were validated as a therapeutic target by using kinase inhibitors selective for the PDGFR-β. PDGFB expressed by tumor cells increases pericytes recruitment in sev-eral *in vivo* models (Guo et al. 1985). Alternatively, genetic abolition of PDGFR-β expressed by embryonic pericytes decreased their recruitment (Abramsson et al. 2003).

In Lewis lung carcinoma implanted in mice, inhibition of endothelial differentiation gene-1 (EDG-1) expression in endothelial cells strongly reduced pericyte coverage (Chae et al. 2004). In a human glioma model developed in rat, Ang-1 led to enhanced pericyte recruitment and increased tumor growth, presumably by favoring angiogenesis (Machein et al. 2004). Alternatively, in a colon cancer model, overexpression of Ang-1 led to smaller tumors with fewer blood vessels and greater pericyte coverage, decreased vascular permeability and reduced hepatic metastasis (Ahmad et al. 2001). Thus, depending on the model, stabilization of blood vessels by Ang-1 may either promote tumor angiogenesis or reduce tumor growth.

In a human neuroblastoma xenotransplanted model, pericyte coverage is halved in tumors grafted on MMP-9-deficient mice and transplantation with MMP-9-expressing bone marrow cells restores the formation of mature vessels (Chantrain et al. 2004). In addition, overexpression of tissue inhibitor of MMP-3 (TIMP-3), a natural inhibitor of MMPs, results in decreased pericyte recruitment in neuroblastoma and melanoma models (Spurbeck et al. 2002). The observation that pericytes express MMPs in many human tumors *in vivo* (Nielsen et al. 1997) suggests that pericytes invasion requires the proteolytic degradation of extracellular matrix by proteases, including MMPs.

VEGF inhibition eliminates tumor vessels without removing pericytes (Morikawa et al. 2002). Anti-angiogenic treatment directed against endothelial cells using VEGF inhibitors induces the regression of tumor vessels and decreased tumor size (Baluk et al. 2005), leading to vessel normalization, characterized by increased pericyte coverage, tumor perfusion and chemotherapeutic sensitivity (Jain 2005). Moreover, removal of VEGF inhibition causes tumor re-growth due to the fact that pericytes provide a scaffold for the rapidly re-growing of tumor vessels (Mancuso et al. 2006).

Pericytes have been indicated as putative targets in the pharmacological therapy of tumors by using the synergistic effect of anti-endothelial and anti-pericyte molecules. Removal of pericyte coverage leads to exposed tumor vessels, which may explain the enhanced effect of combining inhibitors that target both tumor vessels and pericytes. Bergers et al. (2003) showed that combined treatment or pre-treatment with anti-PDGF-B/PDGFBR-β reducing pericyte coverage increases the success of anti-VEGF treatment in the mouse rat insulin promoter-1 T antigen transgene 2 (RIP1-TAG2 model).

However, extensive regression of endothelial cells was not observed in tumors after inhibition of PDGFR-β signaling (Abramsson et al. 2003). STI571 (Gleevec, Imatinib), which targets PDGFRs and other receptor tyrosine kinases, did not reduce vascular density when given alone but did augment the effects of VEGF inhibitors (Bergers et al. 2003). After treatment of RIP1-TAG-2 tumors and Lewis lung carcinomas with AG-013737 or VEGF-Trap, surviving pericytes may become more tightly associated with endothelial cells or have no apparent association with tumor vessels (Inai et al. 2004). Treatment of RIP1-TAG2 tumors with anti-PDGFR-β antibody for 3 weeks reduces pericytes, increases endothelial cell apoptosis but does not seem to reduce tumor vascular density (Song et al. 2005). Similarly, the

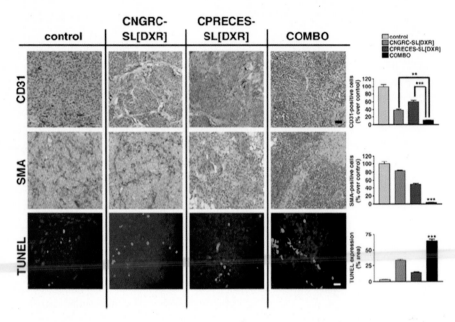

Fig. 2.5 Effects of APN- and APA-targeted SL(DXR) combination on endothelial, perivascular, and tumor cells in vivo. Immunohistochemistry was performed on established neuroblastoma tumors removed from untreated mice (control) or from mice treated with DXR-loaded, APN-targeted or APA-targeted liposomes, or with a combination of the two liposomal formulation (COMBO). Tumors were removed on day 36 and tissue sections were immunostained for CD31 and smooth muscle actin (SMA) to detect endothelial and, respectively, perivascular cells. TUNEL was performed to detect tumor apoptosis. Columns represent mean of CD31, SMA, and TUNEL staining intensities. (Reproduced from Loi et al. 2010)

receptor tyrosine kinase inhibitor SU6668, which also affects PDGFR-β signaling, detaches and reduces pericytes in RIP1-TAG2 and xenotransplanted tumors, thereby restricting tumor growth (Reinmuth et al. 2001; Shaheen et al. 2001).

Sennino et al. (2007) demonstrated that treatment with a novel selective PDGF-B blockade DNA aptamer AX102 that blocks the action of PDGF-B led to progressive reduction of pericytes in Lewis lung carcinomas. More recently, Murphy et al. (2010) generated a series of selective type II inhibitors of PDGFR-β and B-RAF gene targets for pericyte recruitment and endothelial survival, respectively and they demonstrated that dual inhibition of both PDGFR-β and B-RAF exerted synergistic anti-angiogenic activity in both zebrafish and murine models of angiogenesis.

Several other important studies with the aim to target pericytes have been conducted in experimental tumor models (Petras and Hanahan 2005; Maciag et al. 2008; Lu et al. 2010), and even in a human trial in advanced renal cell carcinoma (Haisnworth et al. 2007). Combined targeting of pericytes and endothelial tumor cells with a combination of a peptide ligand of APA, disovered by phage display technology for delivery of liposomal doxorubicin (DXR) to perivascular tumor cells, and aminopeptidase N (APN)-targeted liposomal DXR enhances anti-tumor efficacy of liposomal chemotherapy in human neuroblastoma-bearing mice (Fig. 2.5) (Loi et al. 2010).

Table 2.3 Specific markers for the lymphatic endothelium

VEGFR-3 (transmembrane tyrosine kinase receptor fro VEGF-C and VEGF-D)
Podoplanin (glomerular podocyte membrane glycoprotein)
Prox-1 (homeobox protein required for embryonic lymphatic endothelium)
LYVE-1 (receptor for extracellular matrix/lymph fluid hyaluronan)
Desmoplakin (associates with desmosomal cadherin to form a cell adhesion complex)

2.3 Tumor Lymphatic Vessels

Lymphatic vessels connect peripheral sites, where pathogens first encounter host defenses, with secondary lymphoid tissues, where antigen presentation and the resulting specific immune response originate.

The lymphatic vasculature is composed of a continuous single-cell layer of overlapping endothelial cells that form loose intercellular junctions, and are connected to the extracellular matrix by anchoring filaments. As a consequence of the presence of these elastic anchoring filaments, increased hydrostatic pressure in the tissues causes lymphatic vessels to expand rather than to collapse. With the exception of the epidermis, cornea, and central nervous system, the lymphatic vasculature covers all regions of the body where blood vasculature is also present.

The lymphatic system is comprised of lymphatic capillaries, collecting vessels, lymph nodes and lymphatic ducts. It is important for maintaining homeostasis because it transports excess extracellular fluid and macromolecules from tissues into the systemic circulation, thus conserving a fluid balance in the tissues. Several markers have been discovered that allow the distinction between lymphatic and blood vessels at the capillary level (Table 2.3).

Proliferation of tumor cells in a confined space create mechanical stress, which compresses intra-tumor lymphatic vessels (Padera et al. 2004). Consequently, there are no functional lymphatic vessels in side solid tumors. On the contrary, functional lymphatic vessels are present in the tumor margin and peri-tumoral tissue.

Convincing evidence for tumor lymphangiogenesis has accumulated and show that it can be stimulated in a variety of experimental cancers by VEGF-C and VEGF-D via binding to VEGFR-3 (Cao 2005). VEGF-C overexpression led to lymphangiogenesis and growth of the draining lymphatic vessels, intralymphatic tumor growth and lymph node metastasis in several tumor models (Saharinen et al. 2004) and clinico-pathological studies have reported that expression of VEGF-C, VEGF-D or VEGFR-3 can correlate with lymph node metastasis in human cancer (Stacker et al. 2002). Both lymphangiogenesis and the formation of lymph node metastases can be inhibited by antagonists of VEGF-C and VEGF-D (Cao 2005). Nagy et al. (2002) have demonstrated that in addition to angiogenesis, VEGF-A also induces proliferation of lymphatic endothelium, resulting in the formation of greatly enlarged and poorly functioning lymphatic channels.

He et al. (2004) have shown that lymphatic growth around tumors requires VEGFR-3 signaling. Few lymphatics are present in mice with heterozygous loss of function mutation of VEGF-C, and none are present when VEGFR-3 signaling is blocked.

Isaka et al. (2004) showed that lymphatics around tumors are functionally abnormal, and some may not drain in the correct direction. Skobe et al. (2001) have shown that lymphatic capillaries activated by VEGF-C promote tumor cell invasion by increasing tumor cell transendothelial migration.

Peritumoral lymphatic vessels are larger and more irregular than intratumoral lymphatic vessels, and are more accessible to tumor spreading (Ji et al. 2007). Peritumoral lymphatic vessels are involved in the tumor cell spreading as a consequence of an interstitial fluid hypertension and VEGF-C secretion by tumor cells (Alitalo et al. 2005; Ji 2006). An increased number of peritumoral lymphatic vessels was found in squamous cell carcinoma of the uterine cervix, in lung carcinoma, and in malignant tumors of the endometrium (Gombos et al. 2005; Renyi-Vamos et al. 2005; Adachi et al. 2007; Koukourakis et al. 2005).

Intratumoral lymphatic vessels are small, irregular, with multiple lumens that occasionally contain tumor cells and are less efficient in tumor cell trafficking (Jain and Fenton 2002). Intratumoral lymphatic vessels have been found in squamous cell carcinoma of the head and neck, in malignant melanoma, gastric cancer, pancreatic endocrine tumors and esophageal carcinoma (Beasley et al. 2002; Dadras et al. 2005; Raica et al. 2008; Sipos et al. 2004; Inoue et al. 2008). In breast cancer, some authors reported the absence of intratumoral lymphatic vessels (Vleugen et al. 2004; Williams et al. 2003), while others have demonstrated a highly significant correlation between lymphatic vessels density, lymphovascular invasion and the risk for lymph node metastases (Schoppmann et al. 2004).

Lymphangiogenesis, evaluated as peritumoral and intratumoral lymphatic microvascular density is associated with lymph node metastases in early stage of invasive cervical carcinoma (Zhang et al. 2009) and increased lymphatic microvascular density correlated with lymph node metastasis in early stage invasive colorectal carcinoma (Liang et al. 2006).

The presence of peritumoral and intratumoral lymphatic vessels significantly increases the rate of lymphovascular invasion (Fig. 2.6) (Braun et al. 2008; Marinho et al. 2008). A significant correlation between VEGF-C levels, lymphovascular invasion and lymph node metastases has been reported in cutaneous melanoma (Niakosari et al. 2008; Schmidt et al. 2009). On the contrary, in the tumors of the larynx and hypopharynx it has not been found a correlation between lymphovascular invasion and lymph node metastasis and survival (Volker et al. 2009).

The proliferative activity of tumor-associated lymphatic endothelial cells was demonstrated by double staining with Ki67 and podoplanin/D2-40 in squamous cell carcinoma of the head and neck, melanoma, gastric carcinoma, colorectal carcinoma, non-small lung carcinoma, breast cancer, and squamous cell tumor of the uterine cervix (Fig. 2.7) (Ohno et al. 2007, Dadras et al. 2005; Gao et al. 2009; Omachi et al. 2007; Renyi-Vamos et al. 2005; Koukorakis et al. 2005; Van der Auwera et al. 2005; Mohaanned et al. 2009). On the contrary, in tumors with low lymphangiogenic potential, like renal cell carcinoma, only 2 % of the cases show proliferative lymphatic endothelial cells (Baldewijns et al. 2009).

Tumor associated macrophages influence lymphangiogenesis through NF-kappa B activation (Hagemann et al. 2009). Accumulation of inflammatory cells in tumor

Fig. 2.6 Immunohistochemical expression of D2-40 in intratumoral lymphatic vessels (*arrows*) ins squamous cell carcinoma of the oral mucosa (**a**), and in lymphovascular invasion (**b**). (Reproduced from Raica and Ribatti 2010)

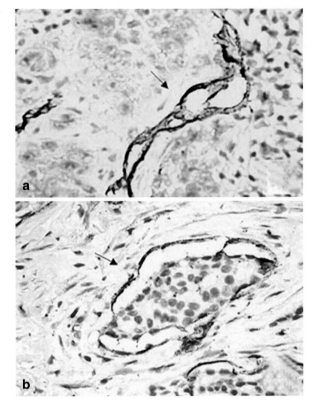

Fig. 2.7 Staining for Ki67/D2-40 in lymphatic endothelial cells in squamous cell carcinoma of the uterine cervix. (Reproduced from Raica and Ribatti 2010)

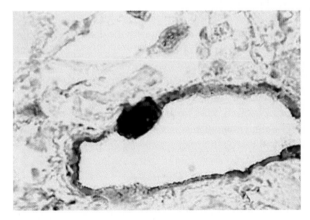

tissues might favor the growth of intratumoral lymphatic vessels, which in turn might promote lymphatic metastasis. The inhibition of inflammatory cells might suppress lymphatic metastasis.

With few exception, all cancers can metastasize. Tumor dissemination involves a series of complex processes including: (i) local invasion into surrounding stromal tissues; (ii) direct seeding of body cavities; (iii) systemic metastasis via tumor-associated blood vessels to distant organs, and (iv) lymphatic metastasis via tumor-associated lymphatic vessels to regional lymph nodes.

Both intratumoral and peritumoral lymphatic vessels are involved in lymph node metastases. VEGF-C and-D promote the metastatic spread of tumor cells through lymphatic vessels (Stacker et al. 2001). Micrometastases to lymph nodes occur before that primary tumors are clinically detectable and a correlation between VEGF-C and -D expression and micrometastases in early gastric cancer has been demonstrated (Arigami et al. 2009). On the other hand, a low frequency of lymph node metastases was found in tumors with low lymphangiogenic activity, like clear-cell renal carcinoma in which only 10 % of the advanced-stage tumors showed intratumoral lymphatic vessels (Baldewijns et al. 2009; Bonsib 2006).

In an experimental model of skin carcinogenesis, lymphangiogenesis within the sentinel lymph nodes mediated by VEGF-A and -C has been demonstrated before metastatic spread (Hirakawa et al. 2005, 2007). Accordingly, in human breast cancer lymphangiogenesis in the sentinel lymph nodes is involved in further lymphatic spread of tumor cells (Van den Eynden et al. 2007) and active lymphangiogenesis was reported in the sentinel lymph node from in small cell lung carcinoma patients mediated by VEGF-A and -C before metastatic spread (Kawai et al. 2008).

It is still unclear how and whether lymph node lymphangiogenesis modifies the lymph node microenvironment an these data indicate that lymph node lymphangiogenesis is an early response for tumor growth and progression.

Gene expression pattern within tumor cells contribute to process of metastasis via the lymphatic system and there are indications that specific changes in gene expression may promote homing of disseminating tumor cells to the lymphatics (Kaifi et al. 2005).

Peritumoral lymphatic microvascular density correlates with lymph node metastasis in squamous cell carcinoma of the uterine cervix (Gombos et al. 2005; Zhang et al. 2009), in lung carcinoma (Sun et al. 2009; Renyi-Vamos et al. 2005; Koukourakis et al. 2005), colorectal carcinoma (Fenzl et al. 2006), in breast cancer (Choi et al. 2005; Schoppmann et al. 2006), endometrial adenocarcinoma (Donoghue et al.2007), and hepatocellular carcinoma (Thelen et al. 2009). In malignant melanoma, some authors found no correlation between lymphatic microvascular density and the risk for lymph node metastasis (Wobster et al. 2006; Sahni et al. 2005), while others considered lymphatic microvascular density as a prognostic marker (Dadras et al. 2005) and useful to discriminate between metastatic and non-metastatic tumors (Shields et al. 2004; Massi et al. 2006).

A significant correlation was found between intratumoral lymphatic microvascular density and lymph node metastases in early invasive carcinoma of the uterine cervix and gastrointestinal tumors (Achen and Stacker 2008; Zhang et al. 2009; Gao et al. 2009). In these latter, lymphatic microvascular density is an independent prognostic factor and correlates with lymph node metastasis, local recurrence and poor outcome of patients (Matsumoto et al. 2006; Kaneko et al. 2006; Chen et al. 2009). Lymphatic microvascular density has an higher predictive value for lymph node metastasis and survival than VEGF-C overexpression (Miyahara et al. 2007; Sugiura et al. 2009).

Podoplanin is expressed by tumor cells of squamous cells carcinoma of the head and neck, lung and uterine cervix (Schacht et al. 2005; Wicki et al. 2006; Kato et al. 2006), malignant mesothelioma (Kimura and Kimura 2005), germ cell tumors (Kato et al. 2005; Mishima et al. 2006a), vascular tumors (Fukunaga et al. 2005), esophageal squamous cell carcinoma (Chuong et al. 2009) gastric cancer (Raica et al. 2008) and some tumors of the central nervous system (Mishima et al. 2006b). Podoplanin favors local invasion through its ability to remodel the actin cytoskeleton of tumor cells, thus increasing their motility (Wicki et al. 2006). A lectin-like specific endogenous receptor of podoplanin was identified on platelets and the interaction with this ligand may regulate tumor invasion and metastases and might be a potential target for therapy (Kato et al. 2007; Suzuki-Inoue et al. 2007). Podoplanin expression was detected mainly in tumor cells at the front of invasion (Wicki and Christofori 2007). An anti-podoplanin antibody inhibited experimental metastases (Kato et al. 2006).

The process of lymph node lymphangiogenesis, induced even before cancer metastasis, indicates that lymphangiogenesis represents a target for the prevention, treatment, and early detection of cancer metastases. VEGFR-3-Ig did not suppress lymph node metastasis when the treatment was started after the tumor cells already entered the lymphatic vessels (He et al. 2005), indicating that complete blockade of lymphatic metastasis require the targeting of both lymphangiogenesis and tumor cell invasion. In an experimental model of human lung tumor cells with high and low metastatic potential, it was found that after the administration of VEGFR-3-Ig, peritumoral lymphatic vessels were scarce in comparison with the control group and incidence of lymph node metastases was reduced. Monoclonal antibodies anti-VEGF-D are efficient in the inhibition of lymphangiogenesis, angiogenesis and metastatic spread (Stacker et al. 2001). Systemic inhibition of VEGFR-3 blocks tumor metastases in the lymph nodes and lung in experimental breast cancer (He et al. 2002; Krishnan et al. 2003). The soluble VEGFR-3 fusion protein (VEGF-C/-D trap) inhibits VEGF-C-induced tumor lymphangiogenesis and metastatic spread in breast cancer xenotransplant model (Karpanen et al. 2001). PTK787 inhibits tumor lymphangiogenesis induced by VEGF-C/-D expression in the Rip1Tag2 mice (Schomber et al. 2009).

VEGF-C siRNA delivered via calcium carbonate nanoparticle or by lentivirus-mediated dramatically suppressed tumor lymphangiogenesis, tumor growth and lymph node metastasis in subcutaneous xenograft of colorectal carcinoma and in breast carcinoma (He et al. 2009; Guo et al. 2009). In an orthotopic breast tumor model it was shown that anti-VEGF-A neutralizing antibody reduces microvascular density and lymphatic microvascular density (Whitehurst et al. 2007). These results suggest a novel mechanism by which anti-VEGF-A therapy suppresses lymphangiogenesis and support its use to control metastasis. Endostatin inhibits lymphangiogenesis and lymph node metastasis by suppressing the production of VEGF-C by tumor cells (Fukumoto et al. 2005) and rapamycin inhibits lymphangiogenesis and lymphatic metastasis *in vitro* and *in vivo* (Kobayashi et al. 2007).

Chapter 3
Mechanisms of Control of Angiogenesis in Tumor Vasculature

3.1 Tumor Angiogenesis

While in the 1990s, tumor angiogenesis was considered as a process with common features, now it has been recognized that tumor type, tumor growth stage, local microenvironment and immunological status, all influence tumor angiogenesis and hence tumor vascular responsiveness to therapeutic intervention.

Angiogenesis and the production of angiogenic factors are fundamental for tumor progression in the form of growth, invasion and metastasis (Fig. 3.1, Ribatti et al. 1999). The process of angiogenesis begins with local degradation of the basement membrane surrounding the capillaries, which is followed by invasion of the surrounding stroma by the underlying endothelial cells, in the direction of the angiogenic stimulus. Endothelial cell migration is accompanied by the proliferation of endothelial cells and their organization into three dimensional structures that join with other similar structures to form a network of new blood vessels.

New vessels promote growth by conveying oxygen and nutrients and removing catabolites (Papetti and Herman 2002). These requirements vary, however, among tumor types, and change over the course of tumor progression (Hlatky et al. 2002). Endothelial cells secrete growth factors for tumor cells and a variety of matrix-degrading proteinases that facilitate tumor invasion (Mignatti and Rifkin 1993). An expanding endothelial surface also gives tumor cells more opportunities to enter the circulation and metastasize (Aznavoorian et al. 1993).

Solid tumor growth occurs by means of an avascular phase followed by a vascular phase. Assuming that such growth is dependent on angiogenesis and that this depends on the release of angiogenic factors, the acquisition of an angiogenic ability can be seen as an expression of progression from neoplastic transformation to tumor growth and metastasis (Fig. 3.1, Ribatti et al. 1999b).

The avascular phase appears to correspond to the histopathological picture presented by a small colony of neoplastic cells that reaches a steady state before it proliferates and becomes rapidly invasive. In this scenario, metabolites and catabolites are transferred by simple diffusion through the surrounding tissue. The cells at the periphery of the tumor continue to reproduce, whereas those in the deeper portion die away. Dormant tumors have been discovered during autopsies of individuals who

D. Ribatti, *Morphofunctional Aspects of Tumor Microcirculation*,
DOI 10.1007/978-94-007-4936-8_3, © Springer Science+Business Media Dordrecht 2012

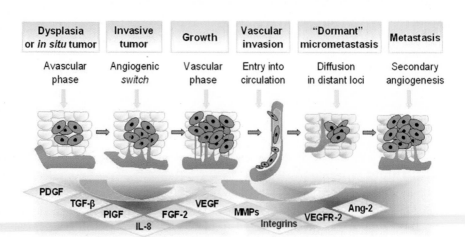

Fig. 3.1 Steps of tumor angiogenesis and growth. (Reproduced from Ribatti and Vacca 2008)

died of causes other than cancer (Black and Welch 1993). Carcinoma *in situ* is found in 98 % of individuals aged 50–70 years who died of trauma, but is diagnosed in only 0.1 % during life. Malignant tumors can grow beyond the critical size of 2 mm at their site of origin by exploiting the host's pre-existing vessels. This occurs in tumors implanted in the rat brain (Holash et al. 1999) and in naturally occurring human lung carcinomas (Pezzella et al. 1997). These finding support the notion that only a very small subset of dormant tumors enters the vascular phase.

Practically all solid tumors, including those of the colon, lung, breast, cervix, bladder, prostate and pancreas, progress through these two phases. The role of angiogenesis in the growth and survival of leukemias and other hematological malignancies has only become evident since 1994 thanks to a series of studies demonstrating that progression in several forms is clearly related to their degree of angiogenesis (Vacca and Ribatti 2006).

Tumor angiogenesis is linked to a switch in the balance between positive and negative regulators, and mainly depends on the release by neoplastic cells of specific growth factors for endothelial cells, that stimulate the growth of the host's blood vessels or the down-regulation of natural angiogenesis inhibitors. In normal tissues, vascular quiescence is maintained by the dominant influence of endogenous angiogenesis inhibitors over angiogenic stimuli (Ribatti et al. 2007b).

The mechanism of this switch was classified by Hanahan, who developed transgenic mice in which the large "T" oncogene is hybridized to the insulin promoter (Hanahan 1985). In this model for β-islet cell tumorigenesis (RIP-Tag model), all islet cells in a transgenic mouse line express the large T antigen at birth. By 12 weeks, 75 % of islets have progressed to small foci of proliferating cells, but only 4 % are angiogenic and their number is closely correlated with the incidence of tumor formation (Hanahan 1985).

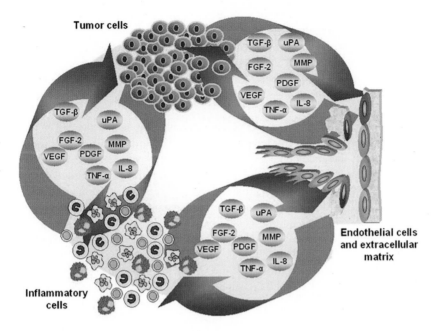

Fig. 3.2 Interplay between tumor cells, inflammatory cells and extracellular matrix in inducing angiogenic response. (Reproduced from Ribatti and Vacca 2008)

The switch depends on increased production of one or more positive regulators of angiogenesis, such as VEGF, fibroblast growth factor-2 (FGF-2), interleukin-8 (IL-8), placental growth factor (PlGF), transforming growth factor-β (TGF-β), PDGF, pleiotrophins and others (Ribatti et al. 2007).

Positive and negative interplay occur between angiogenic growth factors, which can be exported from tumor cells, mobilized from the extracellullar matrix, or released from host cells recruited to the tumor (Fig. 3.2). When VEGF and PlGF are produced by different cells, PlGF homodimers potentiate VEGF-induced angiogenesis and vascular permeability by competing its binding to VEGFR-1 and increasing VEGF availability to VEGFR-2. Upon binding to PlGF, VEGFR-1 could also trans-activate VEGFR-2, leading to enhanced angiogenesis and vascular leakage. In addition, PlGF also binds to neuropilin-1 (NRP-1), to potentiate VEGF function.

It is increasingly recognized that oncogenes, such as mutant RAS or SRC, may also contribute to tumor angiogenesis by enhancing the production of VEGF (Rak et al. 1995; Ellis et al. 1998; Sharma et al. 2005). Down-regulation of the RAS-oncogene in a melanoma driven by doxycycline-inducible RAS led to tumor regression within 12 days (Tang et al. 2005). Cells that expressed low levels of RAS were dormant and non-angiogenic, whereas cells that expressed high levels of RAS developed into full-blown tumors (Watnick et al. 2003). These Authors demonstrated that whereas VEGF levels increased only modestly in tumors that expressed high levels of RAS, TSP-1 levels increased markedly in these cells. Tumors that express bcl-2 escape mitomycin C therapy and grow $\sim$ 1000 mm^3. When bcl-2-expressing tumors are treated with an angiogenesis inhibitor (TNP-470) which selectively inhibits the proliferation of tumor cells or fibroblasts, the bcl-2 effect is annulled, and tumor growth is restricted to <10–15 % of the growth observed in untreated bcl-2 expressing tumors (Fernandez et al. 2001).

Angiogenesis in human tumors is considerably less active than in a physiological condition such as the formation of granulation tissue in the reproductive organs; in fact, the endothelial cell proliferation index value is 0.15 % for the human prostate or breast cancer compared to 6.7 % in granulation tissue and 36 % in the corpus luteum. Moreover, the microvessel densities in human lung, mammary, renal cell and colon carcinomas, glioblastoma and pituitary adenomas are lower than those in their normal counterparts (Eberhard et al. 2002; Turner et al. 2000). In lung carcinoma, for example, the microvessel density was found to be only 29 % that of normal lung tissue. In glioblastoma, microvessel density was found to be 78 % that of normal brain tissue. This apparent paradox is partially explained by the lower oxygen consumption rate of tumor cells (Sterinberg et al. 1997), which are also known to tolerate oxygen deprivation (Graeber et al. 1996). As a result, the intercapillary distance in tumors is greater than in their normal tissue counterparts.

Many areas in the tumor lack vasculature, and existing vessels insufficiently deliver oxygen and remove metabolic products. Consequently, tumor cells suffer from hypoxia, which is a strong stimulus for angiogenesis (Fukumura et al. 2010). In addition, nutrient and serum deprivation leads to the formation of pro-angiogenic reactive oxidative species (Cardone et al. 2005), whereas extracellular acidification, which is a result of high glycolytic metabolism in hypoxic conditions, further enhances angiogenesis (Hunt et al. 2008). There is a complex interrelationship between tumor hypoxia and tumor angiogenesis. Hypoxia in tumors develops in the form of chronic hypoxia, resulting from long diffusion distances between tumor vessels, and/or of acute hypoxia, resulting from a transient collapse of tumor vessels. Many tumors contain a hypoxic microenvironment, a condition that is associated with poor prognosis and resistance to treatment. The production of several angiogenic cytokines, such as FGF-2, VEGF, TGF-β, tumor necrosis factor-α (TNF-α) and IL-8, is regulated by hypoxia. VEGF-mRNA expression is rapidly and reversibly induced by exposure of cultured endothelial cells to low PO$_2$ (Levy et al. 1995). Many tumor cell lines have been reported to show hypoxia-induced expression of VEGF (Papetti et al. 2002; Plate et al. 1992; Shweiki et al. 1992; Potgens et al. 1995; Claffey et al. 1996). In a rat glioma model, VEGF gene expression was activated in a distinct tumor cell subpopulation by two distinct hypoxia-driven mechanisms (Damert et al. 1997). Hypoxia-inducible factor (HIF)-1 helps to restore oxygen homeostasis by inducing glycolysis, erythropoiesis and angiogenesis (Semenza 1996) and tumor vascularization is largely controlled by HIF-1, partly as a result of VEGF upregulation (Carmeliet et al. 1998).

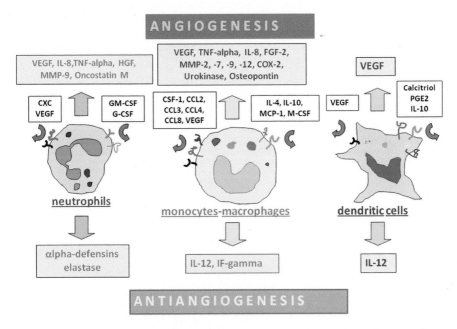

Fig. 3.3 Interplay between angiogenic and anti-angiogenic molecules secreted by neutrophils, monocytes, macrophages, and dendritic cells. (Reproduced from Ribatti and Crivellato 2009)

3.2   The Role of Inflammatory Cells in Tumor Angiogenesis

There is increasing evidence to support the view that angiogenesis and inflammation are mutually dependent (Mueller 2008) During inflammatory reactions, immune cells synthesize and secrete pro-angiogenic factors that promote neovascularization. On the other hand, the newly formed vascular supply contributes to the perpetuation of inflammation by promoting the migration of inflammatory cells to the site of inflammation (Mueller 2008). The extracellular matrix and basement membrane are a source for endogenous angiogenesis inhibitors. On the other hand, many extracellular matrix molecules promote angiogenesis by stabilizing blood vessels and sequestering angiogenic molecules (Nyberg et al. 2008)

It is well established that tumor cells are able to secrete pro-angiogenic factors as well as mediators for inflammatory cells (Ribatti and Vacca 2008). They produce indeed angiogenic cytokines, which are exported from tumor cells or mobilized from the extracellular matrix. As a consequence, tumor cells are surrounded by an infiltrate of inflammatory cells. These cells communicate via a complex network of intercellular signaling pathways, mediated by surface adhesion molecules, cytokines and their receptors (Ribatti et al. 2006). Immune cells cooperate and synergize with stromal cells as well as malignant cells in stimulating endothelial cell proliferation and blood vessel formation (Figs. 3.3, 3.4). These synergies may represent important

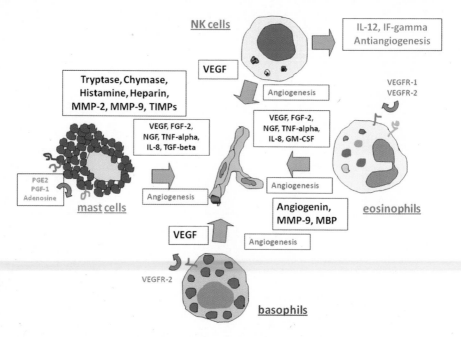

Fig. 3.4 Interplay between angiogenic and anti-angiogenic molecules secreted by NK cells, mast cells, basophils and eosinophils. (Reproduced from Ribatti and Crivellato 2009)

mechanisms for tumor development and metastasis by providing efficient vascular supply and easy pathway to escape. Indeed, the most aggressive human cancers are associated with a dramatic host response composed of various immune cells, especially macrophages and mast cells (Mueller 2008).

3.2.1   Neutrophils

Evidence for the possible role of polymorphonuclear granulocytes in inflammation-mediated angiogenesis and tissue remodeling was initially provided by the finding that CXC receptor-2 (CXCR-2)-deficient mice, which lack neutrophil infiltration in thioglycollate-induced peritonitis (Calcano et al. 1994), showed delayed angiogenesis and impaired cutaneous wound healing (Devalaraja et al. 2000).

During the acute inflammatory response, neutrophils extravasate from the blood into the tissue, where they exert their defence functions. Neutrophils are a source of soluble mediators which exert important angiogenic functions. VEGF, IL-8, TNF-α, hepatocyte growth factor (HGF) and MMPs are the most important activators of angiogenesis produced by these cells (Dubravec et al. 1990; Bazzoni et al. 1991; Grenier et al. 2002). In this perspective, microarray analysis has recently revealed about thirty angiogenesis-relevant genes in human polymorphonuclear granulocytes (Schruefer et al. 2006). Thus neutrophil contribution to pathological angiogenesis may be sustained by an autocrine amplification mechanism that allows persistent VEGF release to occur at sites of neutrophil accumulation. Production and release of VEGF from neutrophils has been shown to depend from the granulocyte colony-stimulating factor (G-CSF) (Ohki et al. 2005). Interestingly, neutrophil-derived VEGF can stimulate neutrophil migration (Ancelin et al. 2004).

In breast cancer, release by tumor-associated and tumor-infiltrating neutrophils of oncostatin M, a pleiotropic cytokine belonging to the IL-6 family, promotes tumor progression by enhancing angiogenesis and metastases (Queen et al. 2005). In addition, neutrophil-derived oncostatin M induces VEGF production from cancer cells and increases breast cancer cell detachment and invasive capacity (Queen et al. 2005). Expression of human papilloma virus (HPV) 16 early region genes in basal keratinocytes of transgenic mice elicits a multi-stage pathway to squamous carcinoma. Infiltration by neutrophils and mast cells, and activation of MMP-9 in these cells coincided with the angiogenic switch in premalignant lesions (Coussens and Werb 1996). In the Rip-Tag2 model of pancreatic islet carcinogenesis, MMP-9-expressing neutrophils were predominantly found in the angiogenic islets of dysplasias and tumors, and transient depletion of neutrophils clearly reduced the frequency of the initial angiogenic switch in the dysplasias (Nozawa et al. 2006). The lack of both MMP-9-positive neutrophils and MMP-2-expressing-stromal cells in mice with a double deficiency for MMP-2 and MMP-9 resulted in a lack of tumor vascularization followed by a lack of tumor invasion (Masson et al. 2005).

Expression of G-CSF or co-expression of G-CSF and granulocyte macrophage-colony stimulating factor (GM-CSF) together induced malignant progression of previously benign factor-negative HaCaT tumor cells. This progression was associated with enhanced and accelerated neutrophil recruitment into the tumor vicinity. The neutrophil recruitment preceded the induction of angiogenesis in the HaCaT heterotransplantation model for human squamous cell carcinoma and in nude mouse heterotransplants of head and neck carcinomas (Obermueller et al. 2004; Gutschalk et al. 2006).

In some tumors, like melanoma, neutrophils are not a major constituent of the leukocyte infiltrate, but they might have a key role in triggering and sustaining the inflammatory cascade, providing chemotactic molecules for the recruitment of macrophages and other inflammatory and stromal cells. Neutrophils produce and release high levels of MMP-9. By contrast, neutrophils secrete little, if any, MMP-2, which plays an important role in the turnover of various extracellular matrix components (Muhs et al. 2003). However, neutrophils release a soluble factor as well as a specific sulphatase and a heparanase that activate latent MMP-2 secreted by other cells and allow releasing of embedded growth factors from the extracellular matrix (Schwartz et al. 1998; Bartlett et al. 1995). Remodeled matrix facilitates the escape of tumor cells leaving the tumor mass to metastasize at distance, because it offers less resistance. In addition, proteolytic enzymes released by neutrophils can diminish cell-cell interactions and permit the dissociation of tumor cells from the original tumor site (Shamamian et al. 2001).

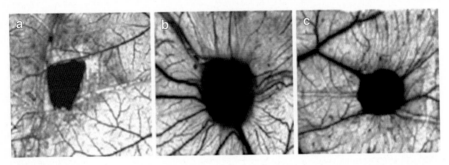

Fig. 3.5 Effects of supernatants of human basophils challenged with buffer (**a**) or with anti-IgE (**b**), on the angiogenic response in the CAM assay. In (**b**), supernatants of anti-IgE-activated basophils induces an angiogenic response, while in (**c**) an anti-VEGF antibody reduces the angiogenic response of anti-IgE-activated basophils supernatants. In (**a**) no vascular reaction is detectable. (Reproduced from De Paulis et al. 2006)

Neutrophils also produce important anti-angiogenic factors. Human neutrophils, for instance, synthesize and secrete small anti-microbial peptides known as alpha-defensins, which exert inhibition of endothelial cell proliferation, migration and adhesion, impaired capillary tube formation *in vitro*, and reduced angiogenesis *in vivo* (Chavakis et al. 2004). In addition, neutrophil-derived elastase can generate the anti-angiogenic factor angiostatin (Scapini et al. 2002) a well known inhibitor of IL-8-, macrophage inflammatory protein (MIP)-2- and growth-related oncogene alpha (GRO-alpha)-induced angiogenesis *in vivo* (Benelli et al. 2002). Remarkably, all-trans retinoic acid, a promising molecule with potential anti-angiogenic use in clinical treatment, has been shown to inhibit VEGF formation in cultured neutrophil-like HL-60 cells (Tee et al. 2006).

3.2.2 Basophils

Basophils express mRNA for three isoforms of VEGF-A and two isoforms of VEGF-B (de Paulis et al. 2006). Peripheral blood and basophils infiltrating sites of chronic inflammation such as nasal polyps contain VEGF-A in their secretory granules. Supernatants of activated basophils induced an angiogenic response *in vivo* in the chick embryo CAM assay (Fig. 3.5). In addition, basophils express VEGFR-2 and NRP-1 which acts as co-receptor for VEGFR-2 and enhances VEGFR-2-induced responses. Remarkably, VEGF-A also functions as basophil chemoattractant providing a novel autocrine loop for basophils self-recruitment (de Paulis et al. 2006). Overall, these data suggest that basophils could play a role in angiogenesis and inflammation through the expression of several forms of VEGF and their receptors. Moreover, basophils release histamine, which displays angiogenic activity in several *in vitro* and *in vivo* settings (Sörbo et al. 1994).

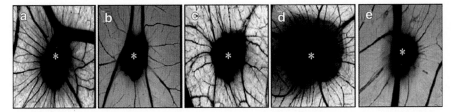

Fig. 3.6 Eosinophils induce an angiogenic response *in vivo* in the CAM assay. In (**a**) eosinophil sonicate; in (**b**) medium alone; in (**c**) cell suspensions of rat mast cells; in (**d**) VEGF; in (**e**) eosinophil sonicate pre-incubated with an anti-VEGF neutralizing antibody, were delivered on the top of the CAM, by using a gelatin sponge implant (*asterisk*). In **a**, **c**, and **d**, gelatin sponges are surrounded by allantoic vessels that develop radially toward the implant in a "spoke-wheel" pattern. In **b**, no vascular reaction is detectable and in **e**, reduced vascular reaction is detectable as compared to **a**. (Reproduced from Puxeddu et al. 2005)

3.2.3   Eosinophils

Eosinophils are pro-angiogenic through the production of an array of cytokines and growth factors, such as VEGF (Horiuchi and Weller 1997), FGF-2 (Hoshino et al. 2001), TNF-α (Wong et al. 1990), GM-CSF (Kita et al. 1991), nerve growth factor (NGF) (Solomon et al. 1998), IL-8 (Yousefi et al. 1995), angiogenin (Hoshino et al. 2001), and are positively stained for VEGF and FGF-2 in the airways of asthmatic patients (Hoshino et al. 2001). Eosinophils release VEGF following stimulation with GM-CSF and IL-5 (Hoshino et al. 2001), both expressed in the tissue and in the bronchoalveolar lavage fluid of patients with allergic asthma. Eosinophils have the capacity to generate VEGF by *de novo* synthesis and to release it (Horiuchi and Weller 1997). Feistritzer et al. (2004) have detected VEGFR-1 and -2 on human peripheral blood eosinophils and they demonstrated that VEGF induces eosinophil migration and eosinophil cationic protein release, mainly through VEGFR-1. Eosinophils promoted endothelial cells proliferation *in vitro* and induce new vessel formation in the aorta ring and in the CAM assays (Fig. 3.6, Puxeddu et al. 2005). Interestingly, neutralization of VEGF in eosinophils reduced their angiogenic effects in the CAM by 55 % suggesting the important, but not unique role played by this factor in the induction of the angiogenic response. Eosinophils are not the only source of VEGF but they can also be targets for VEGF in allergic inflammation. Eosinophil infiltration could be reduced by administration of an anti-VEGF receptor antibody in a murine model of toluene diisocyanate (TDI)-induced asthma (Lee et al. 2002). Because eosinophils are a rich source of preformed MMP-9, it is reasonable to believe that they may promote angiogenesis also by acting on matrix degradation.

Puxeddu et al. (2009) have demonstrated that eosinophil-derived major basic protein (MBP) induced endothelial cell proliferation and enhanced the pro-mitogenic effect of VEGF, but did not affect VEGF release. Moreover, MBP promoted capillarogenesis by endothelial cells seeded on Matrigel and angiogenesis *in vivo* in the CAM assay. Finally, the pro-angiogenic effect of MBP was not due to its cationic charge since stimulation in the CAM with the synthetic polycation, poly-L-arginine did not induce any angiogenic effect (Puxeddu et al. 2009).

3.2.4 Monocytes-macrophages

Cells belonging to the monocyte-macrophage lineage are a major component of the leukocyte infiltration in tumors (Balkwill and Mantovani 2001; Mantovani et al. 2002). A number of tumor-derived chemoattractants ensures macrophage recruitment, including colony-stimulating factor-1 (CSF-1), the CC chemokines CCL2, CCL3, CCL4, CCL5 and CCL8, and VEGF secreted by both tumor and stromal elements (Mantovani et al. 2002). Besides killing tumor cells once activated by interferon-gamma (IFN-γ) and IL-12, tumor-associated macrophages (TAM) produce several pro-angiogenic cytokines as well as extracellular matrix-degrading enzymes (Naldini and Carraro 2005). The stimulating effect exerted by tumor-associated macrophages on the growth of the tumor mass is partly related to the angiogenic potential of these cells. In the tumor microenvironment, macrophages are mainly represented by polarized type II (alternatively activated) or M2 elements, which would derive from tumor-associated macrophages upon local exposure to IL-4 and IL-10 (Mantovani et al. 2002). These cells have poor attitude to destroy tumor cells but are better adapted to promoting angiogenesis, repairing and remodeling wounded or damaged tissues, and suppressing adaptive immunity (Sica et al. 2006). Tumor-associated macrophages represent a rich source of potent pro-angiogenic cytokines and growth factors, such as VEGF, TNF-α, IL-8 and FGF-2 (Ribatti et al. 2007b). In addition, these cells express a broad array of angiogenesis-modulating enzymes, including MMP-2, -7, -9, -12, and cycloxygenase-2 (COX-2) (Sunderkotter et al. 1991; Lewis et al. 1995; Klimp et al. 2001). In humans, a significant relationship between the number of tumor-associated macrophages and the density of blood vessels has been established in tumors like breast carcinoma (Leek et al. 1996), melanoma (Makitie et al. 2001), glioma (Nishie et al. 1999), squamous cell carcinoma of the esophagus (Koide et al. 2004), bladder carcinoma (Hanada et al. 2000), and prostate carcinoma (Lissbrant et al. 2000). In the mouse cornea model, killing of COX-2 positive infiltrating macrophages with clodronate liposomes reduces IL-1-beta-induced angiogenesis and partially inhibits VEGF-induced angiogenesis (Nakao et al. 2005). In one model of subcutaneous melanoma, both angiogenesis and growth rate correlate with tumor infiltration by macrophages that express angiotensin I receptor and VEGF. In addition, Lewis lung carcinoma cells expressing IL-1 beta develop neovasculature with macrophage infiltration and enhance tumor growth in wild-type but not in monocyte chemoattractant protein-1 (MCP-1)-deficient mice, suggesting that macrophage involvement might be a prerequisite for IL-1 beta-induced neovascularization and tumor progression

(Nakao et al. 2005). In a murine model of mammary carcinoma, deficiency of macrophage-colony stimulating factor (M-CSF), a potent inductor of macrophage recruitment in tumor tissues, does not affect early stages of tumor development but reduces progression to invasive carcinoma and metastasis (Lin et al. 2001). This result highlights the possible role of tumor-associated macrophages in contributing to the angiogenic switch that accompanies transition into malignancy. In polyoma middle-T (PyMT)-induced mouse mammary tumors, indeed, focal accumulation of macrophages in premalignant lesions precedes the angiogenic switch and the progression into invasive tumors. Depletion of macrophages in this model severely delayed tumor progression and reduced metastasis, whereas an increase in macrophage infiltration remarkably accelerated these processes. By using the PyMT-induced mouse mammary tumors, Lin et al. (2006) have characterized the development of the vasculature in mammary tumors during their progression to malignancy. They demonstrated that both angiogenic switch and the progression to malignancy are regulated by infiltrated macrophages in the primary mammary tumors. Moreover, inhibition of the macrophage infiltration into the tumor delayed the angiogenic switch and malignant transition, whereas genetic reduction of the macrophage population specifically in these tumors rescued the vessel phenotype. Finally, premature induction of macrophage infiltration into premalignant lesions promoted an early onset of the angiogenic switch independent of tumor progression (Lin et al. 2006).

Depletion of tumor-associated macrophages reduces to about 50 % tumor vascular density, leading to areas of necrosis by loss of blood supply within the tumor mass. Interestingly, macrophages have been shown to accumulate particularly in such necrotic and hypoxic areas in different neoplasia, like human endometrial, breast, prostate and ovarian carcinomas (Ohno et al. 2004; Leek et al. 1999). It is otherwise known, indeed, that up-regulation of the pro-angiogenic program in tumor-associated macrophages, followed by increased release of VEGF, FGF-2, TNF-α, urokinase and MMPs, is stimulated by hypoxia and acidosis (Bingle et al. 2002). Moreover, activated macrophages synthesize and release inducible nitric oxide synthase, which increases blood flow and promotes angiogenesis (Jenkins et al. 1995). Lastly, the angiogenic factors secreted by macrophages stimulate migration of other accessory cells that potentiate angiogenesis, in particular mast cells (Gruber et al. 1995). Osteopontin deeply affects the pro-angiogenic potential of human monocytes (Denhardt et al. 2001). Reports suggest that osteopontin may affect angiogenesis by acting directly on endothelial cells and/or indirectly via mononuclear phagocyte engagement, enhancing the expression of TNF-α and IL-1 ß in mononuclear cells (Leali et al. 2003; Naldini et al. 2006).

It should also be mentioned that monocytes and macrophages are primary producers of IL-12. This multifunctional cytokine can cause tumor regression and reduce metastasis in animal models, due to the promotion of anti-tumor immunity and also to the significant inhibition of angiogenesis (Colombo and Trinchieri 2002). The anti-angiogenic activity is mediated by IFN-γ production, which in turn induces the chemokine IFN-γ-inducible protein-10 (Angiolillo et al. 1995; Romagnani et al.

primarily responsible for promoting angiogenesis. Moreover, although recruited to tumors in lower numbers than TAM, TEM are a more potent source of pro-angiogenic signals, suggesting that they significantly contribute to tumor angiogenesis.

De Palma et al. (2003) proposed that the percentage of incorporated EPC into tumor vessels is very low. TEM, while stimulating angiogenesis, do not actively incorporate into blood vessels and this subpopulation of Tie-2$^+$ cells, rather than bone marrow-derived EPC, which are incorporated in new-forming blood vessels, promote tumor neovascularization through the release of of pro-angiogenic factors.

TEM may also play a role in angiogenesis in wound healing and in non-neoplastic diseases. In mice underwent partial hepatectomy 7–10 days earlier, TEM were found in granulation tissue surrounding the regenerating hepatic lobules, also in proximity of newly-formed vessels, suggesting that they might also contribute to promoting angiogenesis during liver regeneration (De Palma et al. 2003). TEM were not observed in normal tissues suggesting that they may represent a specific subset of resident monocytes (Venneri et al. 2007).

The identification of specific molecules expressed by TEM in tumors could facilitate the design of novel anticancer therapies that selectively target these cells. De Palma et al. (2008), by transplanting hematopoietic progenitors transduced with a Tie-2 promoter/enhancer-driven IFN-α-1 gene, turned TEM into IFN-α cell vehicles that targeted the IFN response to orthotopic human gliomas and spontaneous mouse mammary carcinomas and obtained significant antitumor responses and near complete abrogation of metastasis.

3.2.5 Lymphocytes

Lymphocytes are essential for the airway remodeling. Studies have been performed in mice chronically infected with *Mycoplasma pneumoniae*. Mice lacking B cells expressed a great reduction of angiogenesis when infected with this microorganism (Aurora et al. 2005). The humoral response, indeed, causes deposition of immune complexes on the airway wall, followed by recruitment of inflammatory cells at sites of infected airways which, in turn, are responsible for local production of remodeling factors. Lymphocytes may cooperate to the generation of an anti-angiogenic microenvironment that is essential for causing regression of the tumor mass. For instance, Th cells and cytotoxic T cells are needed to mediate the anti-angiogenic effect of IL-12 (Strasly et al. 2001).

Experimental work suggests that natural killer (NK) cells are required mediators of angiogenesis inhibition by IL-12, and that NK cell cytotoxicity of endothelial cells is a potential mechanism by which IL-12 can suppress neovascularization (Yao et al. 1999). IL-12 receptors indeed are present primarily on NK cells and T cells (Trinchieri 1993). IL-12-activated lymphocytes influence inhibition of tumor growth and function as an anti-vascular agent, by releasing higher level of IFN-γ and down-modulating VEGF (Cavallo et al. 2001).

3.2.6 Dendritic Cells

Dendritic cells are bone marrow, hematopoietic-derived, professional antigen-presenting cells (APC), able to induce both primary and secondary T- and B-cell responses as well as immune tolerance (Bancherau et al. 2000). They participate in the regulation of the inflammatory reaction through the release of cytokines and chemokines (Bancherau and Steinman 2000). Dendritic cells express both pro- and anti-angiogenic mediators when exposed to different combinations of cytokines and microbial stimuli and both positive and negative mediators of the angiogenic process can affect the biology of dendritic cells. Moreover, dendritic cells express both VEGFR-1 and VEGFR-2 (Mimura et al. 2007), and expression of the VEGF co-receptor NRP-1 is induced during *in vitro* differentiation of monocytes into dendritic cells (Bourbié-Vaudanie et al. 2006). Riboldi et al. (2005) reported that dendritic cells can be activated to an angiogenesis-promoting phenotype. They demonstrated that alternative activation of dendritic cells by anti-inflammatory molecules, such as calcitriol, prostaglandin E_2 (PGE$_2$) or IL-10 prompts them to secrete VEGF and inhibit their secretion of IL-12, a potent anti-angiogenic molecule that is secreted by classical activated dendritic cells. Gottfried et al. (2007) demonstrated that incubation of tumor-associated dendritic cells with VEGF and oncostatin M led to transdifferentation into endothelial-like cells. These cells showed strong expression of classical endothelial cell markers, such as vWF and VE-cadherin, while leukocytic markers were reduced. Moreover, they were able to vascular-like tubes on Matrigel.

3.2.7 Mast Cells

Mast cells produce a large spectrum of pro-angiogenic factors. Human, rat and mouse mast cells release preformed FGF-2 from their secretory granules (Qu et al. 1995, 1998a). Human cord blood-derived mast cells release VEGF upon stimulation through FcεRI and c-kit. Both FGF-2 and VEGF have also been identified by immunohistochemistry in mature mast cells in human tissues (Qu et al. 1998b; Grützkau et al. 1998). Human mast cells are a potent source of VEGF in the absence of degranulation through activation of the EP(2) receptor by PGE$_2$ (Abdel-Majid et al. 2004). Following IgE-dependent activation mast cells released several pro-angiogenic mediators stored in their granules, such as VEGF (Boesiger et al. 1998) and FGF-2 (Kanbe et al. 2000), that promote angiogenesis even in the early phase of allergic inflammation. Mast cells can also migrate *in vivo* (Detmar et al. 1998) and *in vitro* (Detoraki et al. 2009) in response to VEGF. Detoraki et al. (2009) have demonstrated that human lung mast cells express VEGF-A, VEGF-B, VEGF-C and VEGF-D at both mRNA and protein level. PGE$_2$ enhanced the expression of VEGF-A, VEGF-B and VEGF-C, whereas an adenosine analog (5'-[N-ethylcarboxamido] adenosine [NECA]) increased VEGF-A, VEGF-C and VEGF-D expression. In addition, supernatants of PGE$_2$- and NECA-activated human lung mast cells induced angiogenic

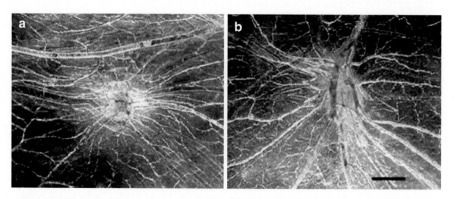

Fig. 3.7 Tryptase and chymase are angiogenic *in vivo* in the CAM assay. In (**a, b**), macroscopic pictures of CAM at day 12 of incubation, treated with tryptase (**a**) and chymase (**b**) respectively. Note the presence of numerous blood vessels converging toward the implant. Original magnification: **a, b**, ×50. (Reproduced from Ribatti et al. 2011b)

response in the CAM assay that was inhibited by an anti-VEGF-A antibody. Finally, PlGF-1 induced mast cell chemotaxis (Detoraki et al. 2009).

Granulated mast cells and their granules, but not degranulated mast cells, are able indeed to stimulate an intense angiogenic reaction in the chick embryo CAM assay. This angiogenic activity is partly inhibited by anti-FGF-2 and -VEGF antibodies, suggesting that these cytokines are involved in the angiogenic reaction (Ribatti et al. 2001). Similarly it has been demonstrated, using the rat-mesenteric window angiogenic assay, that intraperitoneal injection of compound 48/80 causes a vigorous angiogenic response (Norrby et al. 1986). The same treatment in mice also causes angiogenesis (Norrby et al. 1989).

Mast cells store large amounts of preformed active serine proteases, such as tryptase and chymase, in their secretory granules (Metcalfe et al. 1997). Tryptase stimulates the proliferation of endothelial cells and promotes vascular tube formation in culture, is angiogenic *in vivo* in the CAM assay (Fig. 3.7, Ribatti et al. 2011b), and also degrades connective tissue matrix to provide space for neovascular growth. Tryptase also acts indirectly by activating latent MMPs and plasminogen activator (PA), which in turn degrade the extracellular connective tissue with consequent release of VEGF or FGF-2 from their matrix-bound state (Blair et al. 1997). Mast cell-derived chymase degrades extracellular matrix components and therefore matrix-bound VEGF could be potentially released.

Histamine and heparin stimulated proliferation of endothelial cells induced the formation of new blood vessels in the CAM-assay (Ribatti et al. 1987; Sörbo et al. 1994). Histamine stimulates new vessel formation by acting through both H1 and H2 receptors (Sörbo et al. 1994). Heparin may act directly on blood vessels or indirectly by inducing release of FGF-2 from the extracellular storage site. In addition, other cytokines produced by mast cells, such as IL-8 (Moller et al. 1993), TNF-α (Walsh et al. 1991), TGF-β, NGF (Nilsson et al. 1997) and urokinase-type PA have been implicated in normal and tumor-associated angiogenesis (Aoki et al. 2003). Lastly, mast cells also contain preformed MMPs, such as MMP-2 and MMP-9, and TIMPs, which enable mast cells to directly modulate extracellular matrix degradation. This, in turn, allows for tissue release of extracellular matrix-bound angiogenic factors.

Mast cells play a role in tumor growth and angiogenesis. Mast cell-deficient W/W^v mice exhibit indeed a decreased rate of tumor angiogenesis (Starkey et al. 1998). Molecules like heparin could facilitate tumor vascularization not only by a direct pro-angiogenic effect but also through its anti-clotting effect (Theoharides and Conti 2004). In addition, mast cell-derived MMPs can degrade the interstitial tumor stroma and hence release matrix-bound angiogenic factors. An increased number of mast cells has indeed been reported in angiogenesis associated with vascular neo-plasms, like haemangioma and haemangioblastoma (Glowacki and Mullike 1982), as well as a number of solid and haematopoietic tumors. In general, mast cell density correlates with angiogenesis and poor tumor outcome. Association between mast cells and new vessel formation has been reported in breast cancer (Hartveit 1981; Bowrey et al. 2000), colorectal cancer (Lachter et al. 1995) and uterine cervix cancer (Graham and Graham 1966). Tryptase-positive mast cells increase in number and vascularization increases in a linear fashion from dysplasia to invasive cancer of the uterine cervix (Benitez-Bribiesca et al. 2001). An association of VEGF and mast cells with angiogenesis has been demonstrated in laryngeal carcinoma (Sawatsub-ashi et al. 2000) and in small lung carcinoma, where most intratumoral mast cells express VEGF (Imada et al. 2000; Takanami et al. 2000; Tomita et al. 2000). Mast cell accumulation has also been noted repeatedly around melanomas, especially in-vasive melanoma (Reed et al. 1986; Dvorak et al. 1980). Mast cell accumulation was correlated with increased neovascularization, mast cell expression of VEGF (Toth-Jakatics et al. 2000) and FGF-2 (Ribatti et al. 2003a), tumor aggressiveness and poor prognosis. Indeed, a prognostic significance has been attributed to mast cells and microvascular density not only in melanoma (Ribatti et al. 2003b) but also in squamous cell cancer of the oesophagus (Elpek et al. 2001). Angiogenesis has been shown to correlate with tryptase-positive mast cell count in human endometrial cancer and gastric carcinoma (Fig. 3.8), and both parameters were found to increase in agreement with tumor progression (Ribatti et al. 2005b, 2010).

Mast cell density, new vessel rate and clinical prognosis have also been found to correlate in haematological tumors. In benign lymphadenopathies and B cell non-Hodgkin's lymphomas, angiogenesis correlates with total and tryptase-positive mast cell counts, and both increase in step with the increase with malignancy grades (Ribatti et al. 1998, 2000). In non-Hodgkin's lymphomas, a correlation has been found between vessel count and the number of mast cells and VEGF-expressing cells (Fukushima et al. 2001). In the bone marrow of patients with inactive and active multiple myeloma as well as those with monoclonal gammopathies of undetermined significance, angiogenesis highly correlates with mast cell counts (Ribatti et al. 1999a). A similar pattern of correlation between bone marrow microvessel count, total and tryptase-positive mast cell density and tumor progression has been found in patients with myelodysplastic syndrome (Ribatti et al. 2002) and B cell chronic lymphocytic leukemia (Ribatti et al. 2003c). In the early stages of B cell chronic lymphocytic leukemia, the density of tryptase-positive mast cells in the bone marrow has been shown to predict the outcome of the disease (Molica et al. 2003).

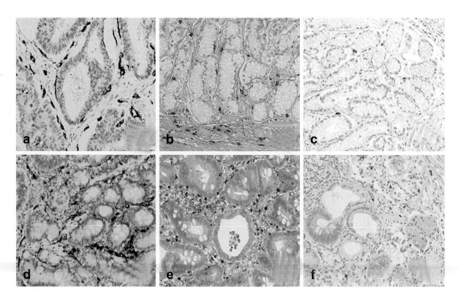

Fig. 3.8 Immunohistochemical staining for CD31, tryptase and chymase in stage II (**a–c**) and stage IV (**d–f**) human gastric cancer. In (**a, d**) endothelial cells immunoreactive for CD31, in (**b, e**) tryptase-positive mast cells; in (**c, f**) chymase-positive mast cells. Blood vessels and mast cells are distributed around the gastric glands. The number of blood vessels and mast cells is higher in stage IV as compared with stage II. Original magnification: **a–f**, ×200. (Reproduced from Ribatti et al. 2010)

3.2.8 Platelets

Human platelets carry in their alpha granules a set of angiogenesis stimulators, such as FGF-2, VEGF and thymidine phosphorylase, and inhibitors, such as endostatin, platelet factor-4 (PF-4) and thrombospondin-1 (TSP-1) (Italiano et al. 2008). These findings may have implications for release of angiogenic molecules at the initiation of wound healing, followed by release of anti-angiogenic molecules at the later stage of wound healing.

These angiogenesis-regulatory molecules are packed into separate and distinc alpha-granules (Italiano et al. 2008). In fact, the treatment of human platelets with a selective proteinase activated receptor-4 (PAR-4) agonist resulted in release of endostatin-containing granules, but not VEGF-containing granules, whereas a selective PAR-1 agonist liberated VEGF, but not endostatin-containing granules (Italiano et al. 2008). Moreover, these molecules are sequestered in platelets in higher concentration than in plasma. In fact, VEGF-enriched Matrigel pellets implanted subcutaneously into mice result in an elevation of VEGF levels in platelets, without any changes in its plasma levels (Klement et al. 2009). Accumulation of platelets in some tumors and release of angiogenic molecules could further stimulate tumor growth. In fact, it has been recently demonstrated that accumulation of angiogenesis regulators in platelets of animals bearing malignant tumors exceeds significantly their concentration in plasma or serum, as well as their levels in platelets from non-tumor bearing mice (Klement et al. 2009). It is likely that novel angiogenesis-regulatory molecules that could be developed into drugs will be discovered in platelets.

3.3   Alternative Mode of Tumor Vascular Growth

Tumor vascular growth may occur almost through other three mechanisms: (i) vasculogenic mimicry (the transdifferentiation of cancer cells allowing them to form tubular structures themselves); (ii) mosaic vessel formation (the incorporation of cancer cells into the vessel wall or vascular cooption); (iii) intussusceptive microvascular growth.

Maniotis et al. (1999) described a new model of formation of vascular channels by human melanoma cells and called it vasculogenic mimicry to emphasize the *de novo* generation of blood vessels without the participation of endothelial cells and independent of angiogenesis. Microarray gene chip analysis of a highly aggressive compared with poorly aggressive human cutaneous melanoma cell lines revealed a significant increase in the expression of laminin 5 and MMP-1, MMP-2, MMP-9 and MT1-MMP in the highly aggressive cells (Seftor et al. 2001), suggesting that they interact with and alter their extracellular environment differently than the poorly aggressive cells, and that increased expression of MMP-2 and MT1-MMP along with matrix deposition of laminin 5 are required for their mimicry.

These data have been disputed by Mc Donald et al. (2000), who consider the evidence presented neither persuasive nor novel. In their opinion, the data are not convincing because three key questions were not addressed: (i) if erythrocytes are used as markers, are they located inside or outside blood vessels?; (ii) where is the interface between endothelial cells and tumor cells in the blood vessel wall?; (iii) how extensive is the presumptive contribution of tumor cells to the lining of blood vessels?

Vasculogenic mimicry appears to be independent of classical angiogenic factors, such as FGF-2 and VEGF, since they are unable to induce a low aggressive melanoma cell line to form tubular networks (Maniotis et al. 1999). Moreover, *in vitro* stimulation with VEGF do not enhance vasculogenic mimicry abilities of these cells (Van der Schaft et al. 2005).

The molecular mechanisms that govern vasculogenic mimicry are still not known in detail. However, VE-cadherin and ephrin A2 seem to be involved, since their single absence hinders cancer cells for tube formation (Hendrix et al. 2001; Hess et al. 2001). More recently, it has been proposed that vasculogenic mimicry could be interpreted as dependent on tumor stem cells (El Hallani et al. 2010; Yao et al. 2011). Vasculogenic mimicry has been reported in Ewing sarcoma (van der Schaft et al. 2005), breast (Shirakawa et al. 2002), ovarian (Sood et al. 2002), prostate

(Sharma et al. 2002), gastric (Li et al. 2010) colorectal, and lung (Passalidou et al. 2002) carcinomas.

Another possibility is that the endothelial cells lining is replaced by tumor cells, resulting in mosaic vessels whose lumen is formed of both endothelial and tumor cells (Chang et al. 2000). These authors used CD31 and CD105 to identify endothelial cells and endogenous green fluorescent protein (GFP) labeling of tumor cells, and showed that approximately 15 % of perfused vessels of a colon carcinoma xenografted at two sites in mice were mosaic, with focal regions where no CD31/CD105 immnoreactivity was detected and tumor cells were in contact with the vessel lumen.

As concerns vascular co-option, Holash et al. (1999) reported that tumor cells migrate to host organ blood vessels in sites of metastases, or in vascularized organs such as the brain, and initiate blood-vessel-dependent tumor growth as opposed to classic angiogenesis. These vessels then regress owing to apoptosis of the constituent endothelial cell, apparently mediated by Ang-2. Lastly, at the periphery of the growing tumor mass angiogenesis occurs by cooperative interaction of VEGF and Ang-2.

Tumor cells often appear to have immediate access to blood vessels, such as when they metastasize to or are implanted within a vascularized tissue (Holash et al. 1999). They immediately co-opt and often grow as cuffs around adjacent vessels. A robust host defense mechanism is activated, in which the co-opted vessels initiate an apoptotic cascade, probably by autocrine induction of Ang-2, followed by regression of the co-opted vessels that carries off much of the dependent tumor and results in massive tumor death. However, successful tumors overcome this regression by neo-angiogenesis.

Shortly after regression, a tumor up-regulates its expression of VEGF, presumably because it is becoming hypoxic due to the loss of vascular support. As in normal vascular remodeling, the destabilizing signal provided by Ang-2, which leads to vessel regression in the absence of VEGF, potentiates the angiogenic response in combination with VEGF. Many solid tumors fail to form a well-differentiated and stable vasculature because their newly formed vessels continue to overexpress Ang-2. Ang-2 induction in host vessels in the periphery of experimental C6 glioma precedes VEGF upregulation on tumor cells, and causes regression of co-opted vessels (Holash et al. 1999).

Vajkoczy et al. (2002) have demonstrated parallel induction of Ang-2 and VEGFR-2 in quiescent host endothelial cells. This suggests that their simultaneous expression is critical for the induction of angiogenesis during vascular initiation of microtumors, and that this, rather than the expression of Ang-2 alone, may indicate the angiogenic phenotype of endothelial cell and thus provide an early marker of activated host vasculature. The VEGF/Ang-2 balance may determine whether the new tumor vessels continue to expand when the ratio of VEGF to Ang-2 is high, or regress when it is low during their remodeling.

Another variant of angiogenesis, different from sprouting, is called intussusceptive microvascular growth (IMG) ("intussusception, known also as or non-sprouting or splitting angiogenesis"); it occurs through the splitting of the existing vasculature by transluminal pillars or transendothelial bridges (Fig. 3.9, Burri and Djonov 2002).

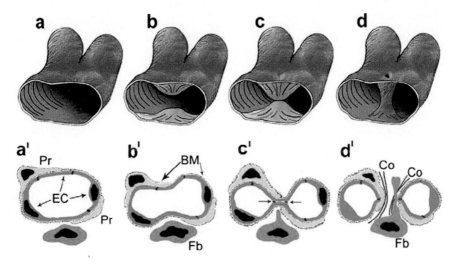

Fig. 3.9 3D (**a–d**) and 2D (**a′–d′**) scheme depicting the generation of transluminarly pillar by intussusceptive angiogenesis. Simultaneously protrusion of opposing capillary walls into the vessel lumen (**a, b; a′, b′**) results in creation of interendothelial contact zone (**c; c′**). In a subsequent step the endothelial bilayer becomes perforated and the newly formed pillar core got invaded by fibroblasts (*fb*) and pericytes (*Pr*), which lay down collagen fibrils (*Co* in **d′**). (Reproduced from Ribatti and Djonov 2012)

It is thought that the pillars then increase in diameter and become a capillary mesh. IMG is a new concept of microvascular growth relevant for many vascular systems, which constitutes an additional, and alternative mechanism to endothelial sprouting. The first reports on IMG were published by Burri and co-workers, who investigated the lung vasculature in postnatal rats and postulated that the capillary network primarily increased its complexity and vascular surface by insertion of a multitude of transcapillary pillars, a process they called "intussusception". Four consecutive steps characterize pillar formation: creation of a zone of contact between opposite capillary walls; reorganization of the intercellular junctions of the endothelium with central perforation of the endothelial bilayer; formation of an interstitial pillar core; subsequently invasion of the pillar by cytoplasmic extensions of myofibroblasts and pericytes, and by collagen fibrils. Lastly, the pillars are thought to increase in diameter and become a capillary mesh (Ribatti and Djonov 2012).

IMG occurs in several tumors, such as colon and mammary carcinomas, melanoma, B-cell non Hodgkin's lymphoma, and glioma (Patan et al. 1996; Djonov et al. 2001b; Crivellato et al. 2003; Ribatti et al. 2005a; Nico et al. 2010). Patan et al. (1996) observed the growth of human colon adenoarcinoma *in vivo* for a period of 6 weeks and demonstrated that IMG is an important mechanism in tumor angiogenesis, suggesting that the rapid vascular remodeling caused by IMG contributes to intermittent blood flow in tumors. Djonov et al. (2001) showed that in mammary tumors of neuT transgenic mice, both sprouting and IMG occur simultaneously in the same nodule. Crivellato et al. (2003) demonstrated that in B-cell non-Hodgkin's lymphomas, large vessels developed transluminal bridges leading to the division of

Fig. 3.10 Ultrastructural pictures of human low grade non Hodgkin's lymphoma specimens. Fusion of intracellular vacuoles causes the formation of large labyrinthic spaces, separate by delicate, branched cytoplasmatic septae and filled by amorphous material. Original magnification, ×8,000. (Reproduced from Crivellato et al. 2003)

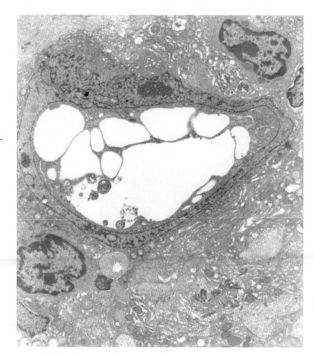

Fig. 3.11 Two examples of tumor vessels with an high a low number of connections of intraluminal tissue folds with the opposite vascular wall, expression of intussusceptive microvascular growth in human primary melanoma tumor specimen with an high in (**a**), and, respectively, low thickness in (**b**). (Reproduced from Ribatti et al. 2005b)

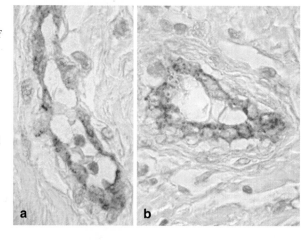

the parental vessels into two or more parts. Moreover, different morphological structural changes in term of expression of IMG have been identified: pillar formation by folding of the lateral vascular wall, fusion of pillars and connection of intraluminal tissue folds with the opposite vascular wall, leading to the splitting of the original vascular structure into newly formed blood vessels (Fig. 3.10). We have confirmed the presence of tumor vessels with connections of intraluminal tissue folds with the opposite vascular wall in human melanoma and glioma (Figs. 3.11, 3.12) (Ribatti et al. 2005b; Nico et al. 2010).

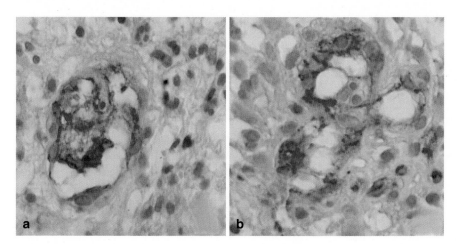

Fig. 3.12 Two examples of tumor vessels with a low and high grade number of connections of intraluminal tissue folds with the opposite vascular wall, expression of intussusceptive microvascular growth in human glioma II in (**a**) and, respectively, IV in (**b**) malignancy grade tumor specimens. (Reproduced from Nico et al. 2010)

Paku et al. (2011) elucidated the mechanism of pillar formation in experimental tumors. By using electron and confocal microscopy, they observed intraluminal nascent pillars that contain a collagen bundle covered by endothelial cells and proposed a new mechanism for the development of pillars consisting of four steps: (i) formation of intraluminal endothelial bridges; (ii) on the abluminal side of the endothelial cells that form the bridge the basement membrane is locally disrupted by proteolytic activity; (iii) an endothelial cell from the bridge adhere to a nearly collagen bundle which is transferred through the lumen, reaches the other side of the lumen, and is transferred into the connective tissue on the other side of the vessel; (iv) further pillar maturation occur through the immigration of fibroblasts/myofibroblasts and pericytes into the pillar and subsequent extracellular matrix proteins (collagen and fibrin) deposition by these cells.

Radiotherapy of mammary carcinoma allografts or treatment with an inhibitor of VEGF tyrosine kinase (PTK787/ZK222854) results in transient reduction in tumor growth rate with decreased tumor vascularization followed by post-therapy relapse with extensive IMG, characterized by a plexus composed of enlarged sinusoidal-like vessels containing multiple transluminal tissue pillars, a dramatic decrease in the intratumoral microvascular density, probably as a result of intussusceptive pruning associated to a minimal reduction of the total microvascular exchange area (Hlushchuk et al. 2008). Moreover, the switch to IMG improves the perfusion of the tumor mass as has been shown by an improvement in oxygen supply of the tumor mass (Hlushchuk et al. 2008).

3.4   The Role of Bone Marrow Stem Cells in Tumor Angiogenesis

The bone marrow compartment comprises the osteoblastic (or endosteal) and the vascular niches. In the endosteal niche, stem cells reside in close proximity to endosteal linings of the bone marrow cavities of the trabecular regions of long bones.

Adult bone marrow contains a population of hematopoietic stem cells (HSC) that can be divided into lineage-positive (Lin$^+$) and lineage-negative (Lin$^-$) categories with regard to their potential to differentiate into formed elements of the blood. Lin$^-$ HSC contain a population of EPC capable of forming blood vessels. HSC dock to spindle-shaped-N-cadherin$^+$ CD45$^-$ osteoblasts and to endothelial cells through a variety of ligand/receptor binding for osteoblasts (OPN/α1-integrin; m-SCF/c-kit; Ang-1/Tie-2; N-cad/N-cad; SDF-1/CXCR-4) and for endothelial cells (VLA-4/V-CAM-1; Tie-2/Ang-2; CXCR-4/SDF-1).

The bone marrow microenvironment switch from a quiescent state to an activated state when local tissue injury results in the release of soluble factors including VEGF, FGF-2, GM-CSF, or osteopontin in the circulation. This in turn promotes the mobilization of both vascular and hematopoietic progenitor cells (HPC) into the peripheral circulation from which these cells are recruited to sites of injury. Mobilization involves the exodus of HSC/HPC from the bone marrow into the circulation. This occurs when stress induces changes of SDF-1 in the bone marrow and is accomplished by the up-regulation of proteases, such as MMP-2, MMP-9, cathepsin-G, and elastase. Upon reaching the bone marrow vasculature, SDF—stimulated HSC/HPC express integrins, such as very late antigen-4(VLA-4) and hyaluronan binding cellular adhesion molecule (CD44). These integrins, in turn, interact with vascular cell adhesion molecule-1 (V-CAM-1), intercellular adhesion molecule-1 (ICAM-1), E- and P-selectins expressed on bone marrow endothelial cells, which slow down the circulating HSC/HSP in a process known as "rolling". Following rolling, firm adhesion and subsequent endothelia transmigration into the hematopoietic compartment is mainly accomplished by VLA-4 interactions. Once extravasated, the cells migrate along extravascular hematopoietic cords toward specific niches through as SDF-1 gradient or reading oxygen gradient originating from the supporting osteoblastic or endothelial niches.

Three antigens have been identified that have helped to distinguish between HC and endothelial cells. CD 146 is expressed almost ubiquitously of endothelial cells but is not found on HC (Bardin et al. 1996; Solovey et al. 2001). In contrast, AC133 is expressed only on primitive HC, and probably on some or all hemangioblasts, but has not been found on endothelial cells (Yin et al. 1997; Miraglia et al. 1997). Finally, CD45 is expressed by most HC, but is not detected on endothelial cells .

Specific populations of bone marrow derived HC contributing to tumor angiogenesis, include GR1$^+$/CD11b$^+$ myeloid progenitors (Yang et al. 2004); F4/80$^+$/CD11$^+$ TAM (Lewis and Pollard 2006); Tie-2$^+$ monocytes (De Palma et al. 2005); CXCR4$^+$/VEGFR1$^+$ hemangiocytes (Jin et al. 2006); CD45$^+$/CD11b$^+$ myeloid cells (Grunewald et al. 2006); PDGFR$^+$ pericyte progenitors (Song et al. 2005); VE-cadherin$^+$/CD45$^+$ vascular leukocytes (Soucek et al. 2007); infiltrating neutrophils and mast cells (Soucek et al. 2007).

3.4.1 Endothelial Precursor Cells

In 1997, Asahara et al. reported the isolation of putative EPC from human peripheral blood, on the basis of cell-surface expression of CD34 and other endothelial markers. These cells differentiate *in vitro* in endothelial cells and seemed to be incorporated at sites of active angiogenesis in various animal models of ischemia. Asahara et al. (1997) used a polyclonal antibody to the intracellular domain of VEGFR-2 to show that when CD34$^+$, VEGFR-2$^+$, CD34$^-$ or VEGFR-2$^-$ cells were injected into mice, rats and rabbits undergoing neovascularization due to hindlimb ischemia, CD34$^+$ and VEGFR-2$^+$ cells, but rarely CD34$^-$ or VEGFR-2$^-$ cells, incorporate into the vasculature.

AC133 is also expressed on EPC subsets, but not on mature endothelial cells (Yin et al. 1997). Its expression is rapidly down-regulated as HPC and EPC differentiate (Yin et al. 1997; Miraglia et al. 1997). Gehlin et al. (2000) demonstrated that AC133$^+$ cells from G-CSF-mobilized peripheral blood differentiate into endothelial cells when cultured in the presence of VEGF and stem cell growth factor. Phenotypic analysis revealed that most of these cells display endothelial features, including the expression of VEGFR-2, Tie-2 and vWF. Peichev et al. (2000) showed that a small subset of CD34$^+$ cells from different hematopoietic sources express both AC133 and VEGFR-2 and incubation with VEGF, FGF-2 and collagen results in their proliferation and differentiation into AC133$^-$, VEGFR-2$^+$ mature endothelial cells.

The majority of circulating EPC reside in the bone marrow in close association with HSC and the bone marrow stroma that provides an optimal microenvironment. Urbich and Dimmeler (2004) define EPC as non-endothelial cells that show clonal expression and stemness characteristics (proliferative capacity and resistance to stress) and are capable of differentiating into endothelial cells. EPC in the peripheral blood may derive from the bone marrow and be not yet incorporated into the vessel wall. In animal models, EPC home in sites of active neovascularization and mobilization of bone marrow-derived EPC, and differentiate into endothelial cells in response to tissue ischemia, vascular trauma or tumor growth (Takahura et al. 1998). Moreover, expansion and mobilization of EPC may augment the resident population of endothelial cells competent to respond to exogenous angiogenic cytokines (Isner and Asahara 1999). Reports on the numeric contribution of EPC to vessel growth are variable, ranging from very low (<0.1 %) to high (up to 50 %) likely dependent on the type of angiogenesis model used (Orlic et al. 2001; De Palma et al. 2003).

One class of EPC is represented by cells that are likely to be hemangioblasts. In mice this includes Sca-1$^+$ c-kit$^+$ lin$^-$ cells and Sca-1$^+$ c-kit$^+$ CD34$^-$ cells from the bone marrow (Jackson et al. 2001; Grant et al. 2002) and in humans, CD113$^+$/CD34$^+$/VEGFR-2$^+$ cells from the bone marrow and blood (Asahara et al. 1997; Peichev et al. 2000; Schatteman et al. 2000; Salven et al. 2003; Nowak et al. 2004). Romagnani et al. (2005) have shown that cells that are positive for CD34/AC133/or CD14 markers also express Nanog and Oct-4, which are embryonic stem cell marker. These cell populations are capable of proliferation in response

Table 3.1 Common markers used to identify human endothelial precursor cells (EPC), mesenchymal stem cells (MSC), hematopoietic stem cells (HSC) and hemapopoietic progenitor cells (HPC)

	EPC	MSC	HSC	HPC
Positive expression	VE-cadherin, vWF, VEGFR-1, VEGFR-2, CD34, CD146, CD105, eNOS, Ulex-europaeus-agglutinin-1 (UEA-1) lectin (binding), acetylated low density lipoprotein (Ac-LDL) (uptake)	CD90, CD44, CD29, CD105, NG2, PDGFRβ	CD34, CD133	Common antigen (CD45); lineage specific markers: myeloid (CD11b, CD14, Ac-LDL (uptake), UEA-1 lectin (binding), CD31, VEGFR-1, VEGFR-2; lymphoid (CD3, CD4, CD8); erythroid (glycophorin A, Ter-119); megakaryocyte (CD41, CD42)
Negative expression	CD45, CD14, CD11b, CD133, CD90	CD45, CD14, CD11b, CD31, VE-cadherin	Lineage-specific markers	CD133

to stem cell growth factors and differentiate into endothelial cells, adipocytes, osteoblasts and neural cells.

Two other different EPC subpopulations have been described, denoted as early and late EPC, with distinct cell growth patterns and ability to secrete angiogenic factors (Gulati et al. 2003; Hur et al. 2004). Early EPC has a monocyte/macrophage phenotype (CD14[+]/CD45[+]/CD68[+] or F4/80[+] in mice), but display endothelial-like markers and behavior and participate in vasculogenic mimicry (Anghelina et al. 2006). Early EPC are spindle-shaped cells, which have a peak growth in culture at 2–3 weeks and which die after 4 weeks and secrete an array of angiogenic, anti-angiogenic and neuroregulatory cytokines (Hur et al. 2004). Late EPC are cobblestone shaped and usually appear after 2–3 weeks of culture, show exponential growth, can be maintained for up to 12 weeks (Gulati et al. 2003; Hur et al. 2004) and are developed exclusively from the CD14[−] (Urbich et al. 2003), CD34[+]/CD45[+] subpopulation (Case et al. 2007).

There is no exclusive EPC marker. It is impossible to differentiate immature EPC from primitive HSC as those cells share common surface, i.e. AC133, CD34 or VEGFR-2 (Table 3.1). Both EPC and mature endothelial cells express similar endothelial-specific markers, including VEGFR-2, Tie-1, Tie-2 and VE-cadherin (Sato et al. 1993; Schnurch and Risau 1993; Vittet et al. 1996; Eichmann et al. 1997). Identification of differences is further complicated by the fact that HC subsets express markers similar to those of endothelial cells, such as CD34, PECAM, Tie-1, Tie-2, Eph and VEGFR-1, transcription factors such as SCL/tal-1 and AML1, vWF (Suda et al. 2000; Lyden et al. 2001). AML1 is expressed in endothelial cells in sites where early HSC emerge, such as the yolk sac (North et al. 1999) and AML1-deficient embryos, which lack definitive hematopoiesis, also display defective angiogenesis in the head and pericardium (Takakura et al. 1998).

When EPC were exposed to angiogenic factors, they formed highly proliferative endothelial colonies, whereas circulating endothelial cells could only generate endothelial monolayers that had limited proliferation capacity because they are mature, terminally differentiated cells. EPC recruitment to sites of neoangiogenesis is triggered by the increased availability of angiogenic growth factors or chemokines, such as VEGF, Ang and SDF-1α (Hattori et al. 2001; Iwaguro et al. 2002; Yamaguchi et al. 2003). The latter binds to the chemokine receptor CXCR-4, which is highly expressed on EPC (Mohle et al. 1998).

Release of EPC from bone marrow into circulation can be included by granulocyte GM-CSF or VEGF and is critically dependent on the activity of endothelial nitric oxide synthase (eNOS) expressed by stromal cells in bone marrow (Aicher et al. 2003). Once avoided at the sites of neovascularization, EPC may recruit additional EPC by releasing growth factors, such as VEGF, HGF, G-CSF and GM-CSF (Rehman et al. 2003). Erythropoietin and estrogen may also positively influence EPC number (Heeschen et al. 2003; Strehlow et al. 2003). Vasa et al. (2001) demonstrated the positive effects of statins on EPC include increasing the number of circulating EPC, reducing senescence, enhancing proliferation rate and differentiation from CD34$^+$ cells.

Angiogenic factors activate MMP-9 (Vu and Werb 2000), which lead to the release of soluble KIT ligand (sKITL) (Engsig et al. 2000) which, in turn, promotes the proliferation and motility within the bone marrow microenvironment, thereby laying the framework for EPC mobilization to the peripheral circulation. Certain inhibitors of tumor neovascularization may act by inhibiting mobilization and homing of EPC to the developing vascular network of tumors. Consistent with this notion, the *in vitro* proliferation and colony-forming ability of human EPC are markedly decreased in the presence of angiostatin (Ito et al. 1999). Kalka et al. (2000) demonstrated that transplantation of EPC to athymic mice with hindlimb ischemia markedly improves blood flow recovery and capillary density in the ischemic limb and significantly reduces the rate of limb loss.

Bone marrow-derived stem cells may be a source of EPC recruited for tumor-induced neovascularization. It has been assumed that the additional endothelial cells required to construct new tumor vessels come from the division and proliferation of local endothelial cells. EPC identity and function in tumor biology has been a matter of controversy due to: (i) the lack of bona fide molecular signature that defines EPC; (ii) conflicting results: the reported fraction of recruited EPC was 90 % in lymphoma models (Lyden et al. 2001), whereas it was a little as 5 % in subcutaneously implanted neuroblastoma; (iii) methodological limitations in discriminating vessel incorporated cells from intimately associated perivascular cells.

EPC have been detected at increased frequency in the circulation of cancer patients and lymphoma bearing mice and tumor volume and tumor secretion of VEGF were found to be correlated with EPC mobilization (Mancuso et al. 2001; Monestiroli et al. 2001). Shaked et al. (2005) demonstrated a strong correlation between tumor growth and EPC numbers in mice using various tumor models and were able to effectively define the optimal anti-angiogenic treatment (anti-VEGFR-2) based on EPC monitoring.

High levels of VEGF produced by tumors may result in the mobilization of bone marrow-derived stem cells in the peripheral circulation and enhance their recruitment into the tumor vasculature (Asahara et al. 1999b; Hattori et al. 2000). Malignant tumor growth results in neoplastic tissue hypoxia, and may mobilize bone marrow-derived endothelial cells in a paracrine fashion and thus contribute to the sprouting of new tumor vessels (Takahashi et al. 1999).

Lyden et al. (2001) demonstrated that transplantation and engrafment of β-galactosidase-positive wild-type bone marrow or VEGF-mobilized stem cells into lethally irradiated Id-mutant mice is sufficient to reconstitute tumor angiogenesis. In contrast to wild type mice, Id-mutants fail to support the growth of tumors because of impaired angiogenesis. The high contribution of EPC in the neovessels of this tumor model results from the fact that recipient Id-deficient mice are angiogenic-deficient as well, and therefore compensatory mechanisms are activated to sustain tumor growth. Tumor analysis demonstrates uptake of bone marrow-derived VEGFR-2$^+$ EPC into vessels surrounded by VEGFR-1$^+$ myeloid cells. Defective angiogenesis in Id-mutant mice is associated with impaired VEGF-induced mobilization and proliferation of the bone marrow precursor cells.

Bertolini et al. (2006) found that when tumor-bearing mice are treated with maximum tolerable dose chemotherapy, there is a marked elevation in EPC counts in peripheral blood during the drug-free break period. Treatment of tumor-bearing mice with vascular disrupting agents (VDA) led to an abrupt release of EPC, which are incorporated into the capillaries of viable peripheral tumor areas (Shaked et al. 2006). Moreover, suppression of EPC mobilization by anti-angiogenic agents resulted in marked reduction in tumor rim size and blood flow as well. Transition from micrometastatic to macrometastatic lesions led to a peak in VEGFR-2$^+$ EPC and EPC could act as the main regulator of the angiogenic switch in progression from micrometastases to macrometastases (Blinder and Fisher 2008).

The percent contribution of EPC to tumor vasculature appears to depend heavily on the experimental model, and thus it remains imprecisely defined. In a bone marrow transplantation model in which donor bone marrow was transduced with a lentiviral vector encoding GFP driven by the endothelial-specific Tie 2 promoter, it was estimated that only 0.05 % of blood vessels in tumor xenografts were derived from bone marrow EPC (Garcia-Barros et al. 2003). Gao et al. (2008) reported by that a very small number of EPC can control the angiogenic switch in a mouse lung metastasis model. Their findings raise the question whether a small number of these angiogenic progenitor cells might have been present (but not detected) in the other transgenic mouse transplant model (Seandel et al. 2008).

Phenotypic definition of bone marrow derived cells and EPC that are present in tumor tissues are difficult to perform especially when the latter cell components are in small numbers. Further elucidation of how the tumor angiogenic process is coordinated resides in a systematic testing and analysis of the biological mediators secreted by the bone marrow derived cells in the various tumor models. The use highly of sensitive bioassay systems such as gene-expression analysis of FACS-purified cells from tumor tissues or the application of proteomic analysis by mass spectrometry or the use of tumor targeted cytokine delivery approach will deliver the promise of elucidating the quantitative and qualitative role each of the bone marrow derived cells play in tumor angiogenesis.

EPC circulate in increased number in the peripheral blood of patients with various types of cancer, including lung (Dome et al. 2006; Pircher et al. 2008), breast (Naik et al. 2008), colorectal (Willett et al. 2005), multiple myeloma (Zhang et al. 2005), myelofibrosis (Massa et al. 2005), non Hodgkin's lymphoma (Igreja et al. 2007), acute myeloid leukemia (Wirzbowska et al. 2005), malignant gliomas (Zheng et al. 2007). In phase I trial with bevacizumab in colorectal cancer the number of EPC decreased (Willett et al. 2005).

3.4.2   Circulating Endothelial Cells

Elevated levels of circulating endothelial cells (CEC) have been found in different types of human malignancies. It is not clear whether or not CEC are simply biomarkers of the accelerated endothelial cells turnover, or are active participants of tumor progression and vascularization.

Lin et al. (2000) found a distinction between vessel wall and bone marrow-derived endothelial cells in blood samples from subjects who had received gender-mismatched bone marrow transplants 5–20 months earlier. They showed that 95 % of CEC had recipient genotype and 5 % had donor genotype. After 9 days of culture, endothelial cells derived predominantly from the recipient vessel wall, expanded only 6-fold, compared with 98-fold after 27 days by endothelial cells, mostly originated from donor bone marrow cells. These data suggest that most CEC in fresh blood originate from vessel walls and have limited growth capability, and that outgrowth of endothelial cells is mostly derived from transplantable marrow-derived cells.

Mancuso et al. (2001) for the first time found that in breast cancer and lymphoma patients both resting and activated CEC were increased significantly. In addition, CEC levels were similar to healthy controls in lymphoma achieving complete remission after chemotherapy, and activated CEC were found to decrease in breast cancer patients after surgery. Beerepoot et al. (2004) reported a significant CEC activation in cancer patients with progressive disease, whereas patients with stable disease had CEC levels comparable to those of health individuals. Zhang et al. (2005) demonstrated an increased number of CEC in multiple myeloma, while Wierzbowska et al. (2005) reported that elevated levels of CEC in acute myeloid leukemia correlated with disease status and response to the treatment. CEC were found to be significantly elevated in breast cancer patients and decreased during chemotherapy (Furstenberger et al. 2006). Rowald et al. (2007) observed that CEC counts were significantly higher in metastatic carcinoma patients compared to healthy controls. Increased CEC levels have been reported in the peripheral blood of patients with gastrointestinal stromal tumor (GIST) (Norden-Zfoni et al. 2007), myelodysplastic syndrome (Della Porta et al. 2007), and chronic lymphocytic leukemia (Go et al. 2008). In a phase I/II study of patients with imatinib-resistant GIST, the Authors found that changes in CEC differed between the patients with clinical benefits and those with progressive disease (Norden-Zfoni et al. 2007).

There is a clear relationship between tumor burdens and EPC/CEC counts in the peripheral blood (Bertolini et al. 2006). Moreover, circulating CEC and EPC counts change with anticancer/anti-angiogenic treatments in preclinical models. A phase II prospective study of low-dose cyclophosphamide given metronomically in combination with celecoxib in adult patients with relapsed or refractory aggressive non-Hodgkin's lymphoma has demonstrated that CEC and EPC declined and remained low in responders (Buckstein et al. 2006). CEC and EPC counts were found to correlate with disease activity and response to thalidomide in patients with multiple myeloma (Zhang et al. 2005). Otherwise, in a phase I study of patients with refractory solid malignancies the differences in the numbers of CEC and EPC between patients and controls were not statistically significant (Twardowski et al. 2008).

3.4.3   Multipotent Adult Progenitor Cells

Reyes et al. (2001) have identified a single cell in human and rodent postnatal marrow that they term bone marrow-derived multipotent adult progenitor cells (MAPC). MAPC were selected by depleting adult bone marrow of HC expressing CD45 and glycophorin-A, followed by long-term culture on fibronectin with EGF and PDGF under low serum conditions. A cell population expressing AC133 and low levels of VEGFR-1 as well as VEGFR-2 and the embryonic stem cell marker Oct-4 emerged. Its culture with VEGF induced differentiation into $CD34^+$, vascular endothelial-cadherin, $VEGFR-2^+$ cells, a phenotype consistent with angioblasts. Subsequently these cells express vWF and markers of mature endothelium, such as CD31, CD36 and CD62-P. They form vascular tubes when plated on Matrigel and up-regulate angiogenic receptors and VEGF in response to hypoxia. In immunocompetent mice, intravenously injected endothelial cells contribute to neovascularization of transplanted tumors and participate in wound healing. Despite extensive efforts, hematopoietic differentiation of human MAPC has not been observed, indicating a significant difference from the pathway by which both HC and endothelial cells are generated from hemangioblasts. Differentiation of human MAPC towards the endothelial lineage was only induced by seeding them at a high density in serum-free medium with the addition of VEGF, whereas culture in medium with foetal calf serum directed the differentiation into osteoblasts, chondroblasts, and adipocytes (Asahara et al. 1999a).

3.4.4   Myeloid-derived Suppressor Cells

Myeloid-derived suppressor cells (MDSC) are a heterogenous population, comprising myeloid progenitors, monocytes and neutrophils, that express low to undetectable levels of MHC-II and co-stimulatory molecules. Yang et al. (2004) found that MDSC obtained from spleens of tumor-bearing mice promoted angiogenesis and tumor growth when co-injected with tumor cells. Reducing the levels of MDSC by either treatment of mice with gemcitabine or by interfering with the Kit ligand/c-Kit receptor axis impaired tumor growth and angiogenesis (Suzuki et al. 2005; Pan et al. 2008). MDSC isolated from tumors of signal transducer and activator of transcription-3 (STAT-3)-deficient mice were markedly less potent in inducing endothelial tube formation *in vitro* as compared to STAT-3 wild-type cells, concomitant with markedly reduced expression levels of several angiogenic factors (Kujawski et al. 2008). These data indicate that while the concept of immature vascular cells delivered to the site of tumor blood vessels was originally developed on bone marrow derived EPC, it is become evident that other classes of vascular cells differentiate from progenitors and adult cells *in situ*.

3.4.5   Mesenchymal Progenitor Cells

The tumor-promoting properties of mesenchymal progenitor cells (MPC) might be related to angiogenesis stimulation. MCP differentiate to form pericytes and are a major source of VEGF secretion (Beckermann et al. 2008). Tumor -associated MCP may secrete pro-inflammatory chemokines which recruit pro-angiogenic hematopoietic cells (Chen et al. 2008). Otherwise, one study suggested that MCP dose-dependently inhibited angiogenesis due to reactive oxygen species (ROS) generation which led to endothelial apoptosis and vessel regression (Otsu et al. 2009). MCP support tumor angiogenesis by providing a supportive role as carcinoma-associated fibroblasts (Haniffa et al. 2009; Mishira et al. 2008) or perivascular mural cells (Au et al. 2008). MPC secrete SDF-1 (Mishira et al. 2008) and are recruited to the site expressing VEGF, PDGF, and EGF (Beckermann et al. 2008).

Chapter 4
Anti-Angiogenesis and Vascular Targeting in Tumor Vasculature

4.1 Anti-Angiogenic Therapy

In 1971, J. Folkman (Fig. 4.1) published in the New England Journal of Medicine a hypothesis that tumor growth is angiogenesis-dependent. The hypothesis predicted that tumors would be unable to grow beyond a microscopic size of 1–2 mm^3 withoutcontinuous recruitment of new capillary blood vessels. Folkman introduced the concept that tumors probably secreted diffusible molecules that could stimulate the growth of new blood vessels toward the tumor and that the resulting tumor neovascularization could conceivably be prevented or interrupted by drugs called angiogenesis inhibitors (Folkman 1971).

Beginning in the 1980s, the biopharmaceutical industry began exploiting the field of anti-angiogenesis for creating new therapeutic compounds for modulating new blood vessel growth in angiogenesis-dependent diseases, and the number of patients receiving anti-angiogenic therapies for cancer treatment has progressively increased.

Anti-angiogenic agents may be classified as synthetic angiogenesis inhibitors and endogenous angiogenesis inhibitors (Table 4.1). Some inhibit endothelial cells directly (blocking endothelial cells from proliferating, migrating or increasing their survival in response to pro-angiogenic molecules), while others inhibit the angiogenesis signaling cascade (blocking the activity of one, two, or a broad spectrum of pro-angiogenic proteins and/or their receptors) or block the ability of endothelial cells to break down the extracellular matrix, which is required to allow endothelial cells to migrate into the surrounding tissues and proliferate into new vessels. Angiogenesis inhibitors may also be characterized by the degree of blocking potential: drugs that block one main angiogenic protein, drugs that block two or three main angiogenic proteins, or drugs that block a range of angiogenic regulators.

4.1.1 Interferons

The antiproliferative activity of IFNs against human tumors was first demonstrated in the 1960s with partially purified IFN-α by Strander at the Karolinska

Fig. 4.1 A portrait of Judah
Folkman, a pioneer in the
study of angiogenesis

Table 4.1 Endogenous
inhibitors of angiogenesis

Matrix derived
Arresten
Canstatin
Endorepellin
Endostatin
Fibronectin fragment (anastellin)
Targeting fibronectin-binding integrins
Fibulin
Thrombospondin-1 and -2
Tumstatin
Non-matrix derived
Growth factors and cytokines
Interferons
Interleukins
Pigment epithelium derived factor (PEDF)
Fragments of blood coagulation factors
Angiostatin
Antithrombin-III
Prothrombin kringle 2
Platelet factor-4
Others
Tissue inhibitors of metalloproteinases (TIMPs)
Chondromodulin
2-methoxyestradiol
Prolactin fragments
Soluble Fms-like tyrosine kinase-1 (S-Flt-1)
Troponin I
Vasostatin

Institute (Strander 1986). A mixture of IFN inhibited the migration of capillary
endothelial cells *in vitro* (Brouty-Boye and Zetter 1980) and lymphocyte-induced
angiogenesis *in vivo* (Sidky and Borden 1987), as well as tumor angiogenesis (Dvorak and Gresser 1989). The first angiogenesis inhibitor IFN-α administered at low

doses was reported in 1980 (Brouty-Boye and Zetter 1980). Since 1988, IFN-α has been used successfully to cause complete and durable regression of life-threatening pulmonary hemangiomatosis, hemangiomas of the brain, airway and liver in infants, recurrent high-grade giant cell tumors refractory to conventional therapy and angioblastomas (Ezekowicz et al. 1992; Kaban et al. 1999; Kaban et al. 2002; Ginns et al. 2003). These tumors all express high levels of FGF-2 as their major angiogenic mediator.

4.1.2   Cartilage and Protamine

Cartilage has been studied as a potential source of an angiogenesis inhibitor because of its avascularity. In fact, cartilage is a relatively tumor-resistant tissue and the tumor associated with cartilage, chondrosarcoma, is the last vascularized of all solid tumors. In 1980, Langer et al. partially purified extracts of cartilage, which inhibited tumor-induced neovascularization when delivered regionally (via controlled release polymer) and when delivered systemically (via infusion) (Langer et al. 1980). Ten years later Moses et al. (1990) purified an angiogenesis inhibitor from bovine scapular cartilage and obtained amino terminal sequence.

In 1982, Taylor and Folkman characterized protamine, a sperm-derived cationic protein, able to inhibit neovascularization in the chick embryo CAM assay and tumor growth and metastases when administered systemically, although its efficacy was limited to its toxicity at high doses (Taylor and Folkman 1982).

4.1.3   Vascular Disrupting Agents

In 1982, Denekamp hypothesized that the local disruption of the tumor vasculature would result in the death of many thousands of tumor cells, and that only a few endothelial cells within the vessels need to be killed to completely occlude the vessels (Denekamp 1982). This strategy relies on the ability of VDA to distinguish the endothelial cells of tumor capillaries from normal ones based on their different phenotype, increased proliferative potential and permeability, and dependence on tubulin cytoskeleton. VDA work best in the poorly perfused hypoxic central tumor areas, leaving a viable rim of well-perfused cancer tissue at the periphery, which rapidly re-grows (Tozer et al. 2005). Combination of VDA and chemo and/or radiation therapy, which targets cancer cells at the tumor periphery, has produced promising responses in preclinical models. By depriving tumors of the nutrients necessary to growth and survive, VDA induce necrosis, particularly within the core of the tumor (Cao 2009). The use of VDA, such as ligand-targeted liposomes and drug conjugates, has started to fulfil its promise. This strategy builds on the clinical success of nanomedicines such as DOXIL®/Caelyx®, which increasing the selective toxicity of chemotherapeutics in cancer by improving therapeutic outcome and/or minimized damage to normal tissues such as heart or bone marrow (Gabizon et al. 1994). Further increases in therapeutic activity can be achieved by using ligand-targeted nanomedicines that have surface-conjugated, tumor-selective antibodies or peptides (Allen 2002), particularly when targeting is by internalizing ligands that facilitate the delivery of the therapeutic contents to intracellular sites of activity via the endosome/lysosome pathway (Allen 2002; Pastorino et al. 2007).

4.1.4   Angiostatic Steroids

When heparin and cortisone were added together in the CAM assay to study angiogenesis activity in fractions being purified from tumor extracts, tumor angiogenesis was inhibited (Folkman et al. 1983). When this combination of heparin and steroid was suspended in a methylcellulose disk and implanted on the CAM, growing capillaries regressed leaving in their place, 48 h later, an avascular zone up to 4 mm in diameter. The anticoagulant function of heparin is not necessary for its angiogenic activity with steroids. In fact, angiostatic steroids administered with heparin fragments that lack anticoagulant activity inhibit angiogenesis. Langer et al. produced a series of heparin fragment, which were tested for their angiostatic activity in the CAM assay (Folkman et al. 1983; Linhardt et al. 1982). A hexasaccharide fragment with a molecular weight of approximately 1,600 was found to be the most potent inhibitor of angiogenesis in the presence of a corticosteroid. Tetrahydrocortisol, a natural metabolite of cortisone is one of the most potent naturally occurring angiostatic steroids. Synthetic angiostatic steroids exhibit greater anti-angiogenic activity than most of the natural steroids. The mechanism of action of angiostatic steroids is not understood completely. However, in the presence of steroid-heparin combinations, the basement membranes of growing capillaries undergo rapid dissolution (Ingber et al. 1986).

4.1.5   Fumagillin

Fumagillin was found by Ingber in the Folkman lab to inhibit endothelial cell proliferation without causing endothelial cell apoptosis, when a tissue culture plate of endothelial cells became contaminated with a fungus *Aspergillus fumigatus fresenius* (Ingber et al. 1990). Scientists at Takeda Chemical Industries (Osaka, Japan) made a synthetic analogue of fumagillin called TNP-470, which inhibits endothelial proliferation *in vitro* at a concentration 3 logs lower that the concentration necessary to inhibit fibroblasts and tumor cells. TNP-470 showed significant inhibition of tumors in clinical trials, including durable complete regression (Milkowski and Weiss 1999). The clinical utility of TNP 470, however, was limited by neurotoxicity. This side effect was overcome when Satchi-Fainaro in the Folkman lab conjugated TNP-470 to HPMA to form caplostatin (Satchi-Fainaro et al. 2005). Caplostatin can be administered over a dose range more than tenfold that of the original TNP-470 without any toxicity. In addition to its anti-angiogenic activity, caplostatin is the most potent known inhibitor of vascular permeability (Satchi-Fainaro et al. 2005).

4.1.6 Thrombospondins

TSP-1 was the first protein to be recognized as a naturally occurring inhibitor of angiogenesis (Good et al. 1990). TSP-1, a heparin-binding protein that is stored in extracellular matrix, was able to inhibit proliferation of endothelial cells from different tissues (Taraboletti et al. 1990) and appeared to destabilize contacts between endothelial cells (Iruela-Arispe et al. 1991). Tumors grew significantly faster in TSP-1 null mice than in wild-type mice (Lawler 2002). Bocci et al. (2003) showed that anti-angiogenic chemotherapy increased circulating TPS-1, and that deletion of TSP-1 in mice completely abrogated the antitumor effect of this anti-angiogenic therapy.

The fact that TSP-1 is a potent endogenous inhibitor of angiogenesis prompted several groups to explore therapeutic applications of TSP-1, endeavouring to identify strategies for up-regulating endogenous TSP-1 and the delivery of recombinant TSP-1 repeats (TSRs) or synthetic peptides that contain sequences from TSRs. Because the whole TSP-1 protein is too large to be used as a therapeutic agent, several mimetic of the anti-angiogenic region have been developed. TSP-2 inhibits endothelial cell migration and tube formation, as well as increasing endothelial cell specific apoptosis through an 80 kDa fragment in the N-terminal region of the molecule (Noh et al. 2003). Various studies have reported results that are also consistent with a pro-angiogenic or biphasic function of TSPs (Nicosia and Tuszynski 1994; Qian et al. 1997; Taraboletti et al. 2000; Chandrasekaean et al. 2000). These findings are consistent with a multicellular nature of TSPs, which enables different domains of the proteins to interact with different receptors of endothelial cells, and thereby elicit very different responses (Bornstein 2001).

4.1.7 Angiostatin

M. O' Reilly came in Folkman lab as postdoctoral fellow in July 1991 and began to test the hypothesis that a primary tumor might suppress growth of its distant metastases by releasing an angiogenesis inhibitor into the circulation by screening a variety of transplantable murine tumors for their ability to suppress metastases. A subclone of Lewis lung carcinoma was isolated that could not suppress metastasis. When the metastasis-suppressing primary tumor was present in the dorsal subcutaneous position, microscopic lung metastases remained dormant at a diameter of less than 200 μm surrounding a pre-existing microvessel, but revealed no new vessels. Within 5 days after surgical removal of the primary tumor, lung metastases became highly angiogenic and grow rapidly, killing their host by 15 days (O'Reilly et al. 1994). This striking evidence demonstrated that the primary tumor could suppress angiogenesis in its secondary metastases by a circulating inhibitor. O'Reilly succeeded in purifying this inhibitor from the serum and urine of tumor-bearing animals and identified a 38 kD internal fragment identical in amino acid sequence to the first four kringle structures of plasminogen, and it was named angiostatin. Angiostatin first revealed that an anti-angiogenic peptide could be enzymatically released from a parent protein that lacked this inhibitory activity (O'Reilly et al. 1994). Angiostatin inhibited growth of primary tumors by up to 98 % (O'Reilly et al. 1996) and was able to induce regression of large tumors and maintain them at a microscopic dormant size.

Endostatin, a 20 kD protein with a M-terminal amino acid sequence identical to the carboxy-terminus of collagen XVIII, provided the first evidence that a basement membrane collagen contained an angiogenesis inhibitor peptide (O'Reilly et al. 1997). O'Reilly in the Folkman lab found endostatin in the blood and urine of mice bearing tumors, which suppressed angiogenesis in remote metastases. Endostatin is a 20–22 kDa C-terminal fragment of type XVIII collagen. It was purified directly from tumor cell-conditioned medium. Endostatin inhibits endothelial cell proliferation and migration, induces apoptosis and causes a G1 arrest of endothelial cells (Dhanabal et al. 1999a, b). The indirect effects on angiogenesis exerted by endostatin include inhibiting of MMP-2 activity, blocking the binding of $VEGF_{165}$ and $VEGF_{121}$ to VEGFR-2, and stabilizing cell-cell and cell-matrix adhesions, preventing the loosening of these junctions required during angiogenesis (Dixelius et al. 2002; Kim et al. 2002). In tumor-bearing animals continuous dosing of endostatin by an intraperitoneal mini-osmotic pump inhibited tumor growth tenfold more effectively than the same dose administered once per day as a bolus dose (Kisker et al. 2001). When endostatin is over-expressed in the vascular endothelium of mice tumors grow 300 % more slowly in mice expressing only 1.6-fold more endostatin that wild-type mice (Sund et al. 2005).

It has been demonstrated that individuals with Down's syndrome have higher levels of circulating endostatin than normal individuals due to an extra copy of the gene for the endostatin precursor on chromosome 21 (Zorik et al. 2001). Interestingly, these subjects have a lower incidence of 200 different human cancers as compared with age-matched controls (Yang et al. 2002). Mice engineering to genetically overexpress endostatin, mimicking individuals with Down'syndrome have slower growing tumors (Sund et al. 2005).

Recombinant endostatin was at first produced in *E. coli*. Preparations of inclusion bodies that were endostatin-free and of low solubility were capable of regressing a variety of established murine tumors when administered subcutaneously (Boehm et al. 1997). When soluble recombinant endostatin was produced in yeats, active endostatin was produced by numerous laboratories and a wide range of inhibited tumors was reported (Folkman and Jalluri 2003). More than 750 reports on endostatin reveal significant inhibition of more than 20 different rat and human tumors (in mice) by administration of the recombinant endostatin protein. Endostatin counteracts virtually all the angiogenic genes up-regulated by either VEGF or FGF-2 and also down-regulates endothelial cell Jun B, HIF-1α, neuropilin and the EGFR (Abdollahi et al. 2004). However, clinical trials using endostatin in cancer patients have yielded only sporadically positive results (Abdollahi et al. 2004).

4.1.8 Thalidomide and Immunomodulatory Drugs

In 1950s, thalidomide was developed as a sedative that showed non toxicity in pre-clinical animal models. In 1962, the association between limb defects in babies born to mothers who used thalidomide during pregnancy was established (Mellin and Katzenstein 1962). D'Amato in the Folkman lab suggested that thalidomide's mechanism of teratogenicity was related to inhibition of angiogenesis in the developing fetal limb bud (D'Amato et al. 1994). Thalidomide has been shown to have pleiotropic effects including anti-angiogenic (down-regulation of TNF-α, FGF-2 and VEGF) immunomodulatory, neurologic, and anti-inflammatory effects (Ribatti and Vacca 2005). Thalidomide was approved in Australia for the treatment of advanced multiple myeloma in 2003 and now is used as a first line therapy. Many patients have been on the drug for 3–5 years without evidence of drug resistance (Ribatti and Vacca 2005).

In terms of milestones in multiple myeloma therapy, the last decade has seen the advent of thalidomide, bortezomib, and lenalidomide. The therapeutic success of these molecules in multiple myeloma treatment is based, at least in part, on their anti-angiogenic activity. We have demonstrated that thalidomide inhibits VEGF, HGF, FGF-2, IGF-1, and Ang-2 expression by multiple myeloma endothelial cells (Fig. 4.2) (Vacca et al. 2005).

A series of immunomodulatory drugs (IMiDs), including lenalidomide and po-malinomide have been developed, which inhibit VEGF and FGF-2 secretion from both myeloma cells and BMSC and block endothelial cell migration and proliferation (Dredge et al. 2002), and demonstrate up to 50,000 times more potent inhibition of TNF-α than thalidomide *in vitro* (Bartlett et al. 2004). We have demonstrated that lenalidomide inhibits angiogenesis and migration of multiple myeloma endothe-lial cells and that lenalidomide-treated multiple myeloma endothelial cells show changes in VEGF/VEGFR-2 signaling pathway, and in several proteins controlling endothelial cell motility, cytoskeleton remodelling, and energy metabolism pathways (Fig. 4.3) (De Luisi et al. 2011).

4.1.9 Bevacizumab

Bevacizumab (Avastin) is a humanized anti-VEGF monoclonal antibody discovered by N. Ferrara at Genentech in San Francisco, California (Fig. 4.4). Because approximately 60 % of human tumors express VEGF-A, Avastin can be very effective against tumors that produce VEGF. Three mechanisms have been proposed for the treatment effects of bevacizumab: (i) an anti-angiogenic mechanism; (ii) the inhibition of circulating endothelial cells and EPC colonizing the tumor vasculature; (iii) its effect on tumor vascular normalization.

Bevacizumab was the first angiogenesis inhibitor approved by the Food and Drug Admnistration (FDA) for the treatment of colorectal cancer in February 2004 (Hurwitz et al. 2004) administered in combination with bolus IFL (irinotecan,

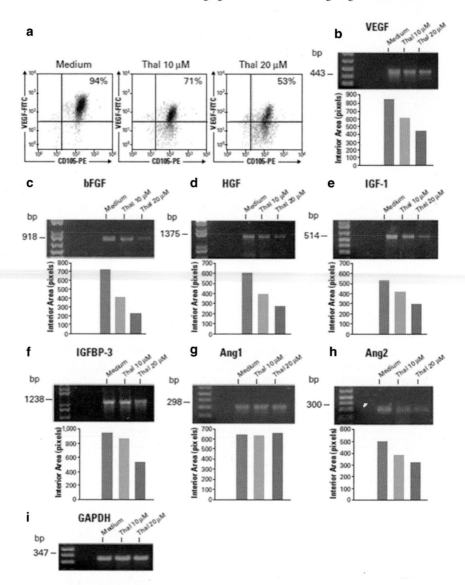

Fig. 4.2 Representative multiple myeloma endothelial cells from a relapsed patients. **a** Fluorescence-activated cell sorting staining with intracellular VEGF and CD105 without (medium) and with exposure to thalidomide (thal) 10 μM and 20 μm. **b–h** RT-PCR gene profile for each cytokine upon the drug exposure is also shown. (Reproduced from Vacca et al. 2005)

5-fluorouracil and leucovorin). This followed for a phase III study showing a survival benefit (Hurwitz et al. 2004). Median survival was increased from 15.6 months in the bolus-IFL plus placebo arm of the trial to 20.3 months in the bolus IFL plus bevacizumab arm. Similar increases were seen in progression-free survival, response

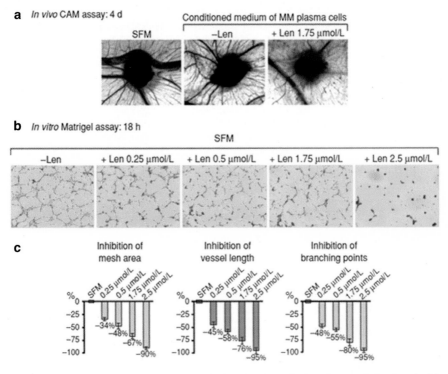

Fig. 4.3 Lenalidomide inhibits angiogenesis in CAM and Matrigel. **a** CAM were incubated with gelatin sponges loaded with SMF (*left*) and with conditioned medium of MM plasma cells either alone (*middle*) or supplemented with lenalidomide (*right*). Note the inhibition of MM angiogenesis by the drug. Original magnification, ×50. **b** Lenalidomide inhibits MMEC angiogenesis in the Matrigel in a dose-dependent manner. MMEC arranged to form a closely knit capillary-like plexus (*left*), whereas the tube formation was gradually blocked with increasing lenalidomide doses. Original magnification, ×80. **c** Skelotonization of the mesh was followed by measurements of its topological parameters: mesh area, vessel length, and branching points. (Reproduced from De Luisi et al. 2011)

Fig. 4.4 A portrait of Napoleone Ferrara, the first to isolate and clone VEGF

rate and duration of response. Bevacizumab has shown encouraging efficacy, increasing survival in patients with human EGFR-2-negative breast cancer or lung cancer when used in combination with chemotherapy (Miller et al. 2007; Sandler et al. 2006), and is currently involved in a broad development program with more than 100 ongoing clinical trials in various indications. Bevacizumab was associated with gastrointestinal perforations and wound healing complications in about 2 % of patients. In addition, the incidence of arterial thromboembolic complications were increased about twofold relative to chemotherapy alone, with patients 65 years or older with a history of arterial thromboembolic events being at higher risk.

Data from clinical trials have shown improved outcomes with the use of bevacizimab as a single agent in metastatic renal cell carcinoma (benefit in progression free survival but not in overall survival) (Yang et al. 2003). In subsequent phase III trials, bevacizumab in combination with standard chemotherapy improved overall survival in lung cancer patients and progression-free survival in breast cancer patients (Jain et al. 2006). Administration of bevacizumab in combination with paclitaxel and carboplatin to patients with NSCLC resulted in increased response rate and time to progression relative to chemotherapy alone in a randomized phase II trial (Johnson et al. 2004).

4.1.10 Tyrosine Kinase Inhibitors

Tyrosine kinase inhibitors (TKI) are small molecules able to pass the plasma membrane (Imai and Takaoka 2006). TKI inhibit not only VEGFR but also other receptors in the superfamily of RTK, including the PDGFR. Inhibitors of VEGF signaling not only interfere with angiogenesis but also cause regression of some tumor vessels (Bergers et al. 2003), causing changes in all components of the vessel wall of tumor, consisting in loss of endothelial cell fenestrations, regression of tumor vessels, and appearance of basement membrane ghosts (Inai et al. 2004). VEGFR TKI generally inhibited or arrested primary tumor growth in mice, but the effects on metastatic process may be different (Hu-Lowe et al. 2008; Padera et al. 2008). Another beneficial strategy is to target VEGF, PDGF, and other receptors with the purpose to inhibit the growth of more blood vessels than with anti-VEGF treatment alone (Erber et al. 2004). Moreover, targeting VEGF/VEGFR system transforms some tumor capillaries into a more normal phenotype (Jain et al. 2006). Finally, VEGF inhibition might have direct cytotoxic effects on tumor cells that aberrantly express VEGFR and depend to some extent on VEGF for their survival. A variety of small-molecules RTK inhibitors targeting the VEGFR have been developed in clinical trials (Table 4.2).

Sorafenib was initially identified as a raf kinase inhibitor and subsequently shown to inhibit several RTK including VEGFR, PDGFR, FMS-like tyrosine kinase 3 (Flt3) and c-Kit. In preclinical models of cancer, sorafenib has shown antiproliferative and anti-angiogenic activity as well as the ability to induce apoptosis in tumor cell lines (Dal Lago et al. 2008; Lee and McCubrey 2003; Panka et al. 2006; Sharma et al. 2005; Wilhelm et al. 2004; Yu et al. 2005). In a Phase II trial in patients with solid tumors, sorafenib monotherapy has shown positive results. Compared with placebo, more patients with renal cell carcinoma treated with sorafenib were progression free at 12 weeks from randomization and median progression-free survival (PFS) was longer

Table 4.2 Kinase inhibitors in clinical trials

AG013736/Axitinib
Brivanib alanitate
AZD2171/Cediranib
Dasatinib
GW654652
Imatinib mesylate
ABT869/Linifanib
AMB706/Motesanib
AE941/Neovstat
Pazopanib/Votrient
BAY43-9006/Sorafenib (NEXAVAR)
SU11248/Sunitinib
SU5416
AV-951/Tivozanib
ZD6474/Vandetanib (ZACTIMA)
PT787/Vatalanib

in the sorafenib group (24 weeks) compared with the placebo group (Ratain et al. 2006). The most common sorafenib-related adverse events (AE) were skin rash, hand–foot reaction and fatigue. The efficacy and safety results have been confirmed in a Phase III trial in which patients with renal cell carcinoma who had previously been treated with one systemic therapy received continuous oral sorafenib or placebo (Escudier et al. 2007). The median PFS in this Phase III study was 5.5 months in the sorafenib group compared with 2.8 months in the placebo group (Escudier et al. 2007) with final results demonstrating an OS advantage in the sorafenib group (Bukowski et al. 2007). Sorafenib has also shown encouraging efficacy and safety results compared to IFN-α in the treatment of patients with renal cell carcinoma (Escudier et al. 2009). Sorefanib monotherapy has also been evaluated in patients with hepatocellular carcinoma. A significantly longer median OS was observed with sorefanib treatment compared with the placebo group (Llovet et al. 2008). The most common AE which were more frequent with sorefanib treatment in this population were diarrhea, weight loss, hand–foot skin reactions and hypophosphataemia. Sorafenib is currently being investigated in combination with multiple other agents, including cytotoxic and targeted agents, in the treatment of several different indications, including renal cell carcinoma, melanoma, hepatocellular carcinoma, and pancreatic, breast, ovarian, thyroid, gastric and colorectal cancers (Dal Lago et al. 2008).

Sunitinib inihibits VEGFR-1, VEGFR-2, PDGFR, c-kit and Flt-3 (Smith et al. 2004). Sunitinib inhibition of tumor angiogenesis with some anti-proliferative and apoptotic effects has been demonstrated in murine xenograft models (Mendel et al. 2003). Sunitinib initially showed efficacy as a second-line therapy in single-arm, Phase II studies in renal cell carcinoma (Motzer et al. 2006, 2007). Patients treated with sunitinib demonstrated an objective response rate of 33 %, a median response duration of 14.0 months, a median PFS of 8.8 months and a median OS of 23.9 months (Motzer et al. 2007). A pivotal Phase III study was then conducted with sunitinib as a first-line treatment in renal cell carcinoma compared with IFN-α (Motzer et al. 2007, 2009), which demonstrated improved OS, PFS and objective response for sunitinib-treated patients compared with those that received IFN-α (Motzer et al. 2009). The most commonly reported sunitinib-related AE were hypertension, fatigue, diarrhea, and hand–foot syndrome. Trials are currently ongoing assessing sunitinib as a second-line therapy (Zimmermann et al. 2009) and in the adjuvant setting (Bellmunt 2009) in the treatment of renal cell carcinoma. Sunitinib has also been evaluated in the treatment of advanced GIST resistant or intolerant to imatinib (Demetri et al. 2006). Median time to progression was 27.3 weeks in patients who received sunitinib compared with 6.4 weeks in the placebo group. Duration of PFS was similar to time to progression. Sunitinib therapy was generally well-tolerated and the most common sunitinib-related AE were fatigue, diarrhea, skin discoloration and nausea. Sunitinib is currently being evaluated in a extensive clinical trial program, alone and in combination with cytotoxic and targeted therapies, in the treatment of multiple indications.

Cediranib (AZD2171) is a potent oral inhibitor of VEGFR, PDGFR and c-Kit which is currently being evaluated in the treatment of glioblastomas, NSCLC, GIST and renal cell carcinoma, amongst other indications. Preclinical studies have established anti-angiogenic and anti-tumorigenic activity in tumor xenograft mouse models with similar results in human tumor xenografts (Wedge et al. 2005; Smith et al. 2007; Gomez-Rivera et al. 2007; Takeda et al. 2007; Goodlad et al. 2006; Miller et al. 2006). Cediranib has also shown favourable results in several Phase I trials in patients with various solid tumors and NSCLC, which has led to the implementation of a clinical trial program evaluating cediranib in colorectal, breast, liver, lung and ovarian cancers, melanoma and mesothelioma (Dietrich et al. 2009).

Encouraging results from a Phase II trial in patients with recurrent glioblastoma were reported (Batchelor et al. 2010) in which patients received open-label cediranib until progression or unacceptable toxicity. The proportion of patients who were progression-free at 6 months was 25.8 % of patients and radiographic partial responses were observed in 56.7 % of patients. Toxicities were reported as manageable and toxicities included hypertension, diarrhea and fatigue. The positive results observed in this Phase II trial are being further investigated with cediranib as a first-line therapy in patients with newly diagnosed glioblastoma. A Phase III trial comparing cediranib monotherapy with cediranib plus lomustine in patients with recurrent glioblastoma is also ongoing (Dietrich et al. 2009). A recent report of a cediranib trial involving 25 patients with refractory or recurrent small cell lung cancer has not been so encouraging, with cediranib failing to demonstrate objective responses and PFS and OS of 2 and 6 months, respectively, at the dose and schedule evaluated (Ramalingam et al. 2010). This suggests that further work is needed to evaluate the potential of cediranib in this indication.

Vatalanib, an oral competitive inhibitor of VEGFR that also targets PDGFR at higher concentrations and has shown clinical activity in Phase I and II trials conducted in patients with several different types of cancers, including NSCLC, glioblastoma and GIST. Results reported for 48 patients with refractory stage IIIB/IV NSCLC, previously treated with platinum-based chemotherapy, are also encouraging (Gauler et al. 2006). A total of 2 % of patients demonstrated a PR, 56 % had SD and 42 % had progressive disease after treatment with vatalanib

(Gauler et al. 2006). In both trials vatalanib was generally well tolerated with nausea amongst the most frequently reported AE in both trials. Results from Phase II trials in patients with metastatic melanoma and metastatic GIST treated with vatalanib have been reported. In thirteen patients with imatinib-resistant GIST, 13 % of patients achieved PR, 53 % had SD and 33 % had progressive disease (Joensuu et al. 2008). The mean time to progression was 8.5 months. As observed in the Phase I trials, vatalanib was generally well tolerated; dizziness, nausea, proteinurea and asthenia, and upper abdominal pain and pyrexia were amongst the most frequently reported AE (Joensuu et al. 2008). In patients with metastatic melanoma treated with vatalanib the tumor control rate was 35 % at 16 weeks with an objective response observed in one patient. Median PFS was 1.8 months and median OS was 6.5 months (Cook et al. 2010). The AE which most commonly required a dose alteration were a combination of Grade 2 toxicities including proteinuria, nausea, fatigue and dizziness (Cook et al. 2010).

Vandetanib (ZD6474) is an oral TKI of VEGFR and EGFR. Doses of $\leq$300 mg vandetanib have been shown to be generally well tolerated with manageable toxicities in patients with solid tumors in both the United States of America and Japan (Holden et al. 2005; Tamura et al. 2006). A series of Phase I and II trials and a Phase III study have gone on to evaluate vandetanib as a monotherapy and in combination with pemetrexed, docetaxel or paclitaxel/carboplatin in patients with advanced NSCLC (Natale et al. 2009; Heymach et al. 2007, 2008; de Boer et al. 2009). In patients with locally advanced or metastatic NSCLC who received vandetanib as second-line therapy to platinum-based chemotherapy, prolonged PFS was observed compared with gefitinib treatment (Natale et al. 2009). Disease control at 8 weeks was achieved in 45 % of patients receiving vandetanib and 34 % in the gefitinib group. A significant advantage in OS was not observed with vandetanib treatment compared with gefitinib treatment (Natale et al. 2009). Adverse events with vandenatib were manageable and included diarrhea, rash and hypertension (Natale et al. 2009). Most recently, disappointing results have been reported for vandetanib in patients with recurrent ovarian cancer with no clinical benefit observed in this patient group (Annunziata et al. 2010), although further evaluation is ongoing.

VEGF-TrapR-1/R-2 (Aflibercept), is a chimeric soluble receptor containing structural elements from VEGFR-1 and VEGFR-2 (Holash et al. 2002). VEGF-TrapR-1/R-2 has shown superior anti-tumor activity compared to other VEGFR-blockers in preclinical models and is currently in clinical trials.

BIBF 1120 is a triple angiokinase inhibitor that simultaneously inhibits VEGFR. PDGFR and FGFR (Hilberg et al. 2008). By inhibiting three receptors that are present on endothelial cells (VEGFR and FGFR), smooth muscle cells (FGFR and PDGFR) and pericytes (PDGFR), BIBF 1120 has the potential to play a critical role in the prevention of tumor growth and spread. BIBF 1120 may be more likely than single-targeted agents to avert the development of resistance, and may provide a treatment option in patients with resistance arising from prior treatment with these agents. *In vitro* studies demonstrated that, out of a panel of 20 other kinases, BIBF 1120 only showed activity for four other members—FLT-3, Src, Lck and Lyn (Hilberg et al. 2008). In preclinical *in vivo* studies, BIBF 1120 produced significant tumor growth inhibition and was efficacious in established tumor xenografts in mice, as shown by dynamic contrast-enhanced MRI (Hilberg et al. 2008). In Phase I clinical studies in patients with advanced solid tumors, BIBF 1120 monotherapy was shown to have a favorable safety profile and be well tolerated at a twice-daily dosing up to the maximum tolerated dose (Okamoto et al. 2010; Stopfer et al. 2011). BIBF 1120 can be administered in combination with carboplatin/paclitaxel, pemetrexed or docetaxel and shows promising efficacy with a consistent safety profile to that of monotherapy (Ellis et al. 2010). Several Phase II studies in various indications have been performed and are also ongoing. A double-blind, two-arm, randomized trial investigating two different doses of BIBF 1120 assessed the efficacy and safety of BIBF 1120 monotherapy in patients with relapsed, advanced NSCLC (Reck et al. 2011). BIBF 1120 was well tolerated, although gastrointestinal toxicities and elevated liver enzymes were observed at a higher frequency in patients receiving BIBF 1120 than those given placebo. Following these promising data, BIBF 1120 has entered Phase III investigation in patients with NSCLC and ovarian cancer.

4.2 Metronomic Chemotherapy

T. Browder in the Folkman laboratory was the first to demonstrate this novel concept: by optimizing the dosing schedule of conventional cytotoxic chemotherapy to achieve more sustained apoptosis of endothelial cells in the vascular bed of a tumor, it is possible to achieve more effective control of tumor growth in mice, even if the tumor cells are drug-resistant (Browder et al. 2000). Browder reported that conventional chemotherapy such as cyclophosphamide administered by the traditional schedule of maximum tolerated doses interspersed with off-therapy intervals of 3 weeks to permit recovery of bone marrow led to a drug resistance in all tumors when therapy was started in Lewis lung carcinomas at tumor volumes of 100–650 mm^3 (Browder et al. 2000).

Conventional chemotherapy is administered at maximum tolerated doses followed by off-therapy intervals of 2–3 weeks to allow the bone marrow and gastrointestinal tract to recover. In contrast, anti-angiogenic chemotherapy is administered more frequently at lower doses, without long interruptions in therapy, and with little or no toxicity. In contrast, when cyclophosphamide was administered at more frequent intervals and at lower doses, it acted as an angiogenesis inhibitor. Proliferating endothelial cells in the tumor vascular bed underwent a wave of apoptosis around 4 days before tumor cell apoptosis began. All tumors completely regressed and animals remained tumor free for their normal lifespan. In an editorial, Hanahan coined the term "metronomic chemotherapy " to indicate the new schedule itself (Hanahan et al. 2000). During anti-angiogenic chemotherapy, endothelial cell apoptosis and capillary dropout precede the death of tumor cells that surround each capillary (Browder et al. 2000). Cyclophosphamide, 5-fluorouracil, 6-mercaptopurine ribose phosphate, and Doxil (the pegylated liposomal formulation of doxorubicin) inhibit angiogenesis when administered on an anti-angiogenic dose schedule.

Kerbel showed that continuous administration of cyclophosphamide in the drinking water inhibited tumor growth in mice by 95 % and significantly increased circulating levels of TSP-1 (Bocci et al. 2003). The low dose "metronomic" chemotherapy was ineffective in TSP-1 null mice, indicating that the low dose oral chemotherapy was in part dependent on its capacity to induce an increase in circulating TSP-1. Pediatric oncologists use a metronomic-like modality of chemotherapies called "maintenance chemotherapy" to treat various pediatric malignancies such as acute lymphoblastic leukemia, neuroblastoma, or Wilm's tumor (Kamen et al. 2006).

4.3 Combinatorial Approach

In addition to using anti-angiogenic agents that target more than one pro-angiogenic factor, another strategy is to use combined modalities. Currently, seven anti-angiogenic agents for treating cancer patients have been approved by FDA, either as single agents or in combination with cytotoxic chemotherapeutics. The abnormalities of the tumor vasculature and the impaired blood flow they cause result in an abnormal microenvironment that is characterized by hypertension, hypoxia and acidosis. These characteristics pose a significant barrier to cancer therapy, with leaky vessels impairing the delivery of therapeutics to the tumor and hypoxia rendering cells resistant to both radiation and many cytotoxic drugs. Therefore, an approach to normalize tumor vessels may correct the tumor microenvironment, making it more susceptible to therapy. Anti-angiogenic therapies have been shown to normalize tumor vasculature and can therefore improve treatment efficacy when co-administered with other therapies (Jain 2005).

Perhaps the most popular combinatorial treatment strategy to date is the co-administration of angiogenesis inhibitors with chemotherapy. A number of *in vitro* studies evaluating the combination of an angiogenesis inhibitor with chemotherapy have been undertaken and the majority report an increased benefit (Fox et al. 2002; Garofalo et al. 2003). Also, the observation that combining chemotherapy with angiogenesis inhibitors causes increased apoptosis in tumors *in vivo* (Inoue et al. 2003; Song et al. 2001; Zhang et al. 2001; Xin et al. 2005) suggests that angiogenesis inhibitors may have an additive effect when administered in combination with chemotherapy.

Clinically, this strategy has also shown encouraging data. In a Phase III trial, bevacizumab in combination with IFL significantly improved survival in patients with previously untreated metastatic colorectal cancer (Hurwitz et al. 2004). Furthermore, a regimen of bevacizumab with chemotherapy has also been shown to improve response and survival of patients with lung cancer (Johnson et al. 2004), breast cancer (Miller et al. 2007) and ovarian cancer (Burger et al. 2010). Encouraging results have also been seen when chemotherapy has been used in combination with cediranib (Goss et al. 2010) and vandetinib (Herbst et al. 2009; de Boer et al. 2009). A combination of cediranib with standard doses of carboplatin and paclitaxel demonstrated encouraging anti-tumor activity, with a significantly higher response rate in NSCLC patients receiving the combination. However, the trial was halted due to intolerability of the 30 mg cediranib dose; an additional trial is ongoing to investigate the efficacy and safety of a 20 mg dose (Goss et al. 2010). In addition, a number of Phase Ib trials are underway evaluating cedarinib in combination with radiotherapy and temozolomide in newly diagnosed glioblastoma (Dietrich et al. 2009), and in combination with cilengitide in recurrent glioblastoma (Dietrich et al. 2009).

As a second-line treatment in patients with advanced or metastatic NSCLC, vandenatib has also shown promising results with improved PFS when given in combination with docetaxel compared with docetaxel alone (Heymach et al. 2007). In combination with paclitaxel/carboplatin in this population, median PFS was similar to that observed in those patients who received paclitaxel/carboplatin alone (Heymach et al. 2008). Similarly, there was no significant difference in OS between these two groups. In both vandenatib combination studies the most common AE included rash, diarrhea and hypertension (Heymach et al. 2007, 2008). Comparable prolongation of PFS and OS was observed in a trial investigating vandetenib in combination with pemetrexed; however, neither reached statistical significance (de Boer et al. 2009).

An alternative approach is to use a combined-modality strategy of an anti-angiogenic agent with radiation, a logical treatment strategy given that tumor progression is a major reason why radiotherapy fails and these agents inhibit such progression (Horsman and Siemann 2006). Furthermore, the degree of intratumoral hypoxia, an important determinant of response to radiotherapy, is influenced by angiogenesis inhibitors (Pang and Poon 2006). Preclinical *in vitro* studies have demonstrated that this strategy has been associated with an enhanced response, which was seen regardless of whether an intermittent or concomitant (Geng et al. 2001; Griscelli et al. 2000; Hess et al. 2001; Huber et al. 2005; Abdollahi et al. 2003) schedule was used. However, there have been few clinical studies completed to date that combine anti-angiogenic agents with radiotherapy. One retrospective study has demonstrated that the combination of sorafenib with palliative radiotherapy increased PFS and was well tolerated in patients with metastatic renal-cell carcinoma (Kasibhatla et al. 2007). Anti-tumor activity has also been reported in a Phase I study of bevacizumab added to fluorouracil- and hydroxyurea-based concomitant chemo-radiotherapy in patients with head and neck cancer (Seiwert et al. 2008), although a dose reduction of chemotherapy was required due to neutropenia.

The final promising strategy that we will consider is combining targeted agents. There are two concepts of combination-targeted therapy: horizontal blocking, whereby numerous targets downstream from a target are individually or jointly inhibited; or vertical blocking, whereby the same signaling pathway is targeted at two different levels. Horizontal blocking generally involves the simultaneous inhibition of angiogenesis, tumor cell proliferation and promotion of tumor cell apoptosis. The vertical blocking strategy is exemplified by the combination of sorafenib and bevacizumab, which simultaneously target the VEGF pathway (Sosman et al. 2007). The combination of bevacizumab and sunitinib has also shown anti-tumor activity with a tolerable side-effect profile in a Phase I trial in patients with advanced solid tumors (Rini et al. 2009), although another study of this combination in patients with

Table 4.3 The most frequently used carrier systems for tumor vasculature targeted drug delivery

Peptides
Growth factors
Antibodies and fragments
Modified polymers, proteins
Modified virus
Modified liposomes
Modified immune effector cells

renal-cell carcinoma has resulted in high levels of toxicity, often leading to study discontinuation (Feldman et al. 2009). Horizontal blocking has also been shown to be effective in a study in patients with NSCLC—the concomitant administration of sorafenib with gefitinib (an EGFR TKI) was seen to be well tolerated and promising signs of efficacy were observed (Adjei et al. 2007). This combination has also shown impressive anti-tumor activity in a Phase I study in patients with metastatic renal-cell carcinoma, based on both objective response and time to progression, although there was some increased toxicity, including increased incidence of hand–foot syndrome (Sosman et al. 2008). Clearly, the option of combining targeted agents is an attractive idea that has shown encouraging results, but more studies are required to ensure that an increased burden of toxicity does not outweigh the benefits of increased efficacy.

4.4 Vascular Targeting to the Tumor Vasculature

Endothelial cells lining tumor blood vessels express several cell surface markers that are absent in quiescent blood vessels. Ligand-directed vascular targeting can be accomplished by antibodies, specific peptides or growth factors complexed with immunomodulatory, procoagulant or cytotoxic molecules (Table 4.3) (Thorpe 2004). Fusion proteins and chemical conjugates of VEGF and diphtheria toxin or gelonin induced tumor regression in mice (Thorpe 2004). Retroviruses have been engineered so that thay can be coated with an antibody (e.g., anti-VEGFR-2) for the selectively delivery of genes to tumor endothelium (Masood et al. 2001).

The phage display library technique has been successfully used to discover tumor cell surface binding peptides, that may alleviate some of the problems associated to antibody targeting. A panel of peptide motifs, including the sequences CDCRGDCFC (termed RGD-4C), NGR, CPRECES and GSL, have been assembled that target the tumor blood vessels (Kolonin et al. 2001).

Since endothelial cells in angiogenic vessels express several proteins that are absent or barely detectable in established blood vessels, it should be possible to develop ligand-targeted chemotherapy strategies, based on peptides that are selective for tumor vasculature. These proteins include α_v integrins (Nemeth et al. 2007), receptors for angiogenic growth factors (Rafii and Lyden 2008), and other types of membrane-spanning molecules, such as the APN and APA (Sato et al. 2007; Marchiò et al. 2004). *In vivo* panning of phage libraries in tumor-bearing mice have proven useful for selecting peptides that bind to receptors that are either over-expressed or are selectively expressed on tumor-associated vessels and that home to neoplastic tissues (Sergeeva et al. 2006).

Among the various tumor-targeting ligands identified to date, peptides containing the asparagine-glycine-arginine (NGR) motif, which binds to CD13, have proven useful for delivering various anti-tumor compounds to tumour vasculature (Pastorino et al. 2003a; Curnis et al. 2002). Although there are several subpopulations of CD13 (O'Connell et al. 1991), relatively widely distributed in the body, only one isoform is believed to be the receptor for the NGR-containing peptides, exclusively expressed in angiogenic vessels, such as the neovasculature found in tumor tissues (Colombo et al. 2002). Consequently, since the CD13 isoform that is recognized by NGR-containing peptides is expressed on endothelial cells within most solid tumors, an alternative strategy that has been pursued to increase the delivery of anti-cancer/anti-angiogenic compounds such as DXR to tumors is based on the use of vascular-targeted liposomes.

There are several advantages of targeting chemotherapeutic agents to proliferating endothelial cells in the tumor vasculature rather than directly to tumor cells: (i) the tumor vasculature endothelial cells are partially independent of the type of solid tumor; hence, the destruction of proliferating endothelial cells in the tumor microenvironment can be effective against a variety of malignancies; (ii) acquired drug resistance, resulting from genetic and epigenetic mechanisms, reduces the effectiveness of available drugs (Klement et al. 2000). The strategy of using tumor vasculature-targeted liposomal drugs, has the potential to overcome problems of tumor cell heterogeneity and drug resistance, since the tumor vasculature is derived from local and circulating endothelial cells that are genetically stable; (iii) anti-vasculature therapies may also circumvent what may be a major mechanism of intrinsic drug resistance, namely insufficient drug penetration into the interior of a tumor mass due to high interstitial pressure gradients within tumors (Jain 1998); (iv) the fact that a large number of cancer cells depend upon a small number of endothelial cells for their growth and survival might also amplify the therapeutic effect (Jain 2001); (v) oxygen consumption by neoplastic and endothelial cells, along with poor oxygen delivery, creates hypoxia within tumors, which compromises the delivery and effectiveness of conventional cytotoxic therapies as well as molecularly targeted therapies (Jain 1998, 2001), but vasculature targeting could still be effective in these tumors.

Anti-vasculature strategies are reported to leave a cuff of unaffected tumor cells at the tumor periphery that are independent of the tumor vasculature, and that can subsequently re-grow and kill the host (Huang et al. 1997). Further therapeutic improvements might be possible by using a combined strategy that targets the tumor vasculature, and, using a different targeting agent and/or drug, targets the residual tumor cells (Pastorino et al. 2006).

In the development and optimization of anti-vasculature and anti-tumor therapies, the importance of choosing the correct animal models cannot be underestimated. Most preclinical studies of tumor angiogenesis and of anti-angiogenic therapy employ rapidly growing transplantable mouse tumors or human tumor xenografts, which are grown as solid tumors localized to the subcutaneous space, and this approach almost certainly exaggerates the anti-tumor responses to experimental therapies. Unlike in the clinic, distant metastases are usually not the focus of the treatment, but it is

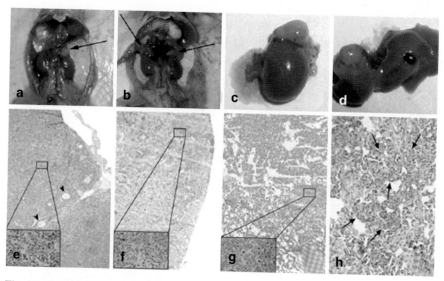

Fig. 4.5 Orthotopic neuroblastoma xenograft model in SCID mice. **a** and **b**, adrenal gland tumors (*arrows*) in mice that were injected orthotopically with neuroblastoma cells at 14 (**a**), and 21 (**b**) days before sacrifice. **c** and **d**, representative right adrenal gland (**c**), and liver (**d**) samples at 3 and 4 weeks after injection of neuroblastoma cells, respectively. **e–h**, histological analysis of representative ovary (**e**), kidney (**f**), liver (**g**), and lung (**h**) samples. Forty days after cell injection, animals were sacrificed, the organs were removed, fixed, paraffin embedded, sectioned and stained with hematoxylin & eosin. *Arrows* indicate metastatic tumor invasion in the lung. *Arrowheads* show the normal ovaric follicular structure surrounded by tumor neuroblastoma cells. (Reproduced from Pastorino et al. 2003a)

precisely such secondary tumors which are ultimately responsible for cancer's lethality. To elucidate possible influences of the host microenvironment, angiogenesis-specific studies of tumors have been carried out in an orthotopic location (Pastorino et al. 2003b). The use of orthotopically transplanted tumors may be preferable, not only to induce or enhance the incidence of metastases, but also because the response of a tumor mass growing ectopically may be abnormal compared with the same tumor growing in a physiologically relevant site (Moss et al. 1991; Fidler 1995; Fidler and Ellis 1994). Thus, the optimal xenograft model of neuroblastoma would be in an orthotopic site at an appropriate development stage, designed to mimic the environment of endogenous neuroblastoma (Fig. 4.5) (Pastorino et al. 2006; Khanna et al. 2002). Using an orthotopic murine neuroblastoma model, Pastorino et al. (2003b, 2006) have shown neuroblastoma tumor regression, pronounced destruction of tumor vasculature and increased life span, when the mice were treated with doxorubicin-loaded liposomes targeted via a surface grafted a NGR-containing peptide that binds to, and is internalized by, the angiogenic endothelial cell marker APN (Fig. 4.6). Pharmacokinetic studies in the same model indicated that systemically administered liposomes coupled to NGR peptide had long-circulating profiles in blood. They were cleared only slightly faster than non-targeted formulations; approximately 30 % of both drug and carrier remained in circulation at 24 h after administration. The uptake

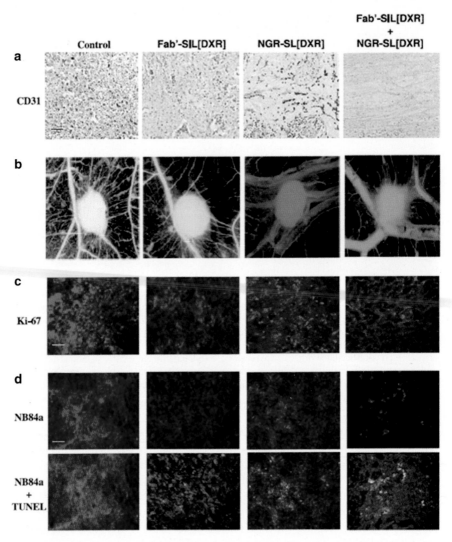

Fig. 4.6 Effects of combination therapies on angiogenesis, tumor cell proliferation, and apoptosis *in vivo*. **a**, **c**, and **d**, immunohistochemical analysis of neuroblastoma metastases removed from untreated mice or mice treated with individual or combined liposomal formulations. Tumors were harvested on day 50 and tissue sections were immunostained for (**a**) CD31, (**c**) Ki-67, (**d**) NB84a to show neuroblastoma cells or else double-labeled for NB84a and TUNEL, to detect tumor apoptosis. **b**, angiogenesis inhibition in vivo in the CAM model for combinations of Fab′-SIL(DXR) and NGR-SL(DXR). (Reproduced from Pastorino et al 2006)

of the NGR-targeted liposomes into neuroblastoma tumors was at least three times higher than that of non-targeted liposomes after 24 h, and the doxorubicin content of the liposomes could be observed spreading outside the blood vessels into the tumors. Five percent of both the liposomes and the drug had localized to tumor by

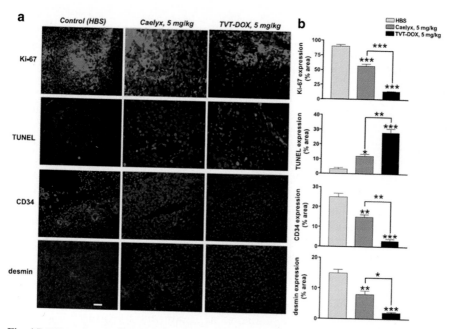

Fig. 4.7 Effect of doxorubicin-contained formulations on tumor cell proliferation, apoptosis and endothelial tumor cell suppression. **a** Histologic analysis were done at day 29 on primary lung cancer removed from untreated (HBS) mice and from mice treated at 7 day after cell inoculation with one intravenous inicetion doxorubicin per week ×5 week at a doxorubicin dose of 5 mg/kg/dose in either Caelyx or TVT-DOX. **b** Morphometric assessment of Ki-67, TUNEL, CD34, and desmin expression. (Reproduced from Pastorino et al. 2008a)

24 h post-injection, and this had increased to about 13 % by 48 h post-injection. No tumor uptake was observed for liposomes coupled with a control peptide (Pastorino et al. 2003b).

Pastorino et al. (2008a, b) validated the potential of the anti-vascular strategy, by evaluating tumor vasculature (NGR)-targeted liposomal doxorubicin (TVT-DOX) in several murine xenografts of doxorubicin-resistant human cancers, including lung, ovarian and neuroblastoma. Compared to an untargeted formulation of DOXIL®/Caelyx®, which is in clinical use for the treatment of ovarian cancer and other solid tumors (Northfelt et al. 1997; Gordon et al. 2000), the GMP preparation of TVT-DOX was able to more effectively kill angiogenic tumor blood vessels and, indirectly, the tumor cells that these vessels support (Fig. 4.7). Moreover, the anti-tumor activity of TVT-DOX was higher than that of DOXIL®/Caelyx®, in all three solid tumor murine models, particularly when administered at the higher dose. This suggests that TVT-DOX should be evaluated as a novel VTA/VDA strategy for adjuvant therapy of solid tumors.

In order to assess the effect of TVT-DOX in controlling minimal residual disease (MRD) and in helping to prevent tumor relapse, a new neuroblastoma model was set up to take advantage of new imaging techniques. Briefly, mice were orthotopically injected with luciferase-transfected neuroblastoma cells on day 0, and half of the mice had their tumors surgically resected on day 20, (Pastorino et al. 2007). The resected neuroblastoma animal model, referred as "neuroblastoma-resected", was used to visualize, by bioluminescene imaging (BLI) and MRI evaluation, the response of MRD to therapy, after surgical removal of the primary mass. As well, the model was used to monitor orthotopic expansion over time, and organ-specific homing of tumor cells (Pastorino et al. 2008b). Interestingly, a comparison of the results obtained from each of the two different imaging devises was important in evaluating the effect of the TVT-DOX therapy in controlling primary tumor growth, relapse and minimal residual disease.

Images were evaluated for BLI intensity over time. Treatment of neuroblastoma-resected mice with TVT-DOX induced a partial arrest in primary tumor re-growth and possibly an inhibition of MRD, in four of five treated mice, while images from control (buffer) neuroblastoma-resected mice showed tumor mass relapse and expansion in four of five mice. In neuroblastoma-resected mice receiving TVT-DOX treatment, there was an increased life span compared to controls, with two of five animals still alive at 130 days after tumor challenge (Pastorino et al. 2008b). In some experiments, the visualization of MRD in neuroblastoma-resected mice, and the response to TVT-DOX therapy, were quantified by a 1.5 T magnetic resonance scanner before and after gadolinium injection (Pastorino et al. 2008b). Images were obtained of the coronal and axial planes, perpendicular to the vertebral column of the animal. MRI was performed before and after i.v. injections of gadolinium.

To be highlighted is the fact that BLI and MRI are viable real-time, non-invasive, quantitative and qualitative methods for monitoring the response to VDA therapy. Each has its advantages. While BLI was more sensitive than MRI for detecting early tumor responses to therapy, MRI was able to precisely identify focal MRD and appears to be more suited to the study of tumor growth *in vivo* and the effects of chemotherapy in experimental animal models (Pastorino et al. 2008b).

In the above described studies, APN-targeted liposomes showed an enhanced anti-tumor and angiostatic effect against all the tumor animal models examined to date, and is a candidate for progression to clinical investigation. In the last years, nanomedicine has become a rapid growth research area (Torchilin 2005), particularly for anticancer applications. Several nanomedicines, primarily lipid-based drug carriers such as DOXIL®/Caelyx® (Gabizon et al. 2008), have received clinical approval, and several more lipid-based and polymeric carriers are undergoing clinical evaluation (Allen et al. 2006). A logical extension of this success is to further improve the anti-tumor effects of liposomal nanomedicines, in a more selective manner, through the use of "active-targeting moieties", coupled to their external surface (Allen 2002; Pastorino et al. 2008b). Receptor-mediated internalization of nanomedicines into tumor cells is mandatory for improved therapeutic efficacy of targeted-liposomal drugs (Pastorino et al 2003b; Sapra and Allen 2002). It has also been established that nuclear localization of DOX is required for cell killing, as it binds DNA and inhibits topoisomerase II (Taatjes and Koch 2001). Liposomal DOX was able to enter into the cells and localize in the nucleus when it was targeted via the NGR-containing peptide coupled at the external surface of the nanoparticles, but not in the absence of the targeting agent (Garde et al. 2007). However, clinical trials based on the use of single, either pro- or anti-vasculature molecules can be more challenging than anticipated, and monotherapy with single angiogenesis or anti-vasculature inhibitors might not be sufficient to control cancer and the myriad of angiogenic factors produced by cancer cells. Phage display bio panning on viable cells is a powerful approach for identifying cell-specific peptides that mediate binding to individual tumor types (Elayadi et al. 2007). This technology, based on the principle that bacteriophages can present specific binding ligands on their surface, has been used for discovering peptides that can specifically bind to organs, tumors, or cell types (Sergeeva et al. 2006; Pasqualini and Ruoslahti 1996).

In the future, it will be possible to use phage display techniques on tumor patient specimens in order to develop novel ligand-targeted liposomal chemotherapeutic strategies that are based on the selective targeting of other novel molecular markers, expressed on the tumor vasculature or the tumor cell surface itself. Thus, a multiple target approach, based on a combination of anti-tumor and anti-vascular therapies, analogous to combination chemotherapy currently in widespread clinical use, could be expected to improve the therapeutic effects of nanomedicine drugs against many types of adult and pediatric solid tumors.

4.5 Tumor Vascular Normalization

Excess production of pro-angiogenic factors and/or diminshed production of anti-angiogenic molecules may be considered responsible of the vascular structure anomalies in tumors (Ribatti et al. 2007c). Restore of this balance may induce a normalization of structure of blood vessels. The concept of "normalization" of tumor blood vessels by anti-angiogenic drugs was introduced by Rakesh Jain in 2001. The state of normalization is probably transient, and dependent on the dose and duration of the treatment.

VEGF inhibition could temporarily eliminate the immature, ineffective vessels in the tumor vasculature, restore or normalize the function of tumor-associated vasculature, decreasing vascular permeability in conjunction with restoration of sustained pressure gradients, as demonstrated by intravital imaging studies in preclinical models and in cancer patients Fukumura et al. (2010), thereby enhancing systemic delivery of oxygen or perfusion of cytotoxic agents to intratumoral sites (Tong et al. 2004; Jain 2005). Moreover, abrogation of VEGF signaling increases collagenase IV activity, leading to restoration of normal basement membrane (Winkler et al. 2004), which generally in tumors has an abnormally thickness (Baluk et al. 2003).

Temporal vascular nomalization induces vascular regression, which causes tumor hypoxia (Sato 2011). Hypoxia could alter the property of cancer cells through the induction of HIF-1, as HIF-1 is reported to be involved in the induction of genes that induce invasive and metastatic properties of tumor cells (Semenza 2003). Hypoxia generated by angiogenesis inhibition triggers pathways that make tumors more aggressive and metastatic and less sensitive to antiangiogenic treatment, as demosntrated by Paez-Ribes et al. (2009), who used blocking VEGFR-2 antibodies to mouse models of pancreatic neuroendocrine carcinoma and glioblastoma, and found that cancers showed heightned invasiveness or metastasis.

Pericytes of tumor vessels have an abnormally loose association with endothelial cells, have an irregular shape and cytoplasmic projections into the tumor parenchyma (Morikawa et al. 2002). Irregularities in their coverage correlate with increased vascular permeability and less effective vascular delivery (Ribatti et al. 2011). Pericytes are believed to protect the remaining vessels and defend against the anti-VEGF treatment (Bergers et al. 2003). Blockade of VEGFR-2 has been shown to increase total pericyte vasculature. In fact, blockade of VEGF signaling promotes pericyte recruitment in Lewis lung carcinomas, RIP-TAG-2 tumors (Inai et al. 2004) and other tumor models (Tong et al. 2004; Willett et al. 2004), by triggering increased tumor production of Ang-1 (Winkler et al. 2004). Surviving blood vessels of RIP-TAG-2 tumors treated with anti-VEGF agents have a reduced expression of VEGFR-2 and VEGFR-3 in addition to a more normal caliber (Ribatti et al. 2011). Helfrich et al. (2010) demonstrated in spontaneously developing melanomas of MT/ret transgenic mice after using anti-VEGF therapy and in human melanoma metastases taken at clinical relapse in patients undergoing adjuvant treatment with bevacizumab, that tumor vessels which are resistant to anti-VEGF therapy are characterized by enhanced vessel diameter and normalization of the vascular bed by coverage of mature pericytes.

Bevacizumab has been approved by the FDA for treatment of recurrent glioblastoma (Norden et al. 2008, 2009), prolongs progression-free survival and controls peritumoral edema, but its effects on overall survival remain to be determined. The decrease of brain edema due to vascular normalization is thought to be an important factor of its benefit (Martin et al. 2009). Other inhibitors of VEGF, VEGFRs and other proangiogenic signaling pathways are being evaluated. MRI scans showed that treatment with cediranib (AZ 2171), a small molecule inhibitor of VEGFR, lowered blood vessel size and permeability, consistent with the hypothesis of vascular normalization (Batchelor et al. 2007). Cediranib monotherapy was used in a clinical trial for recurrent glioblastoma with encouraging radiographic response and 6-month progression-free survival (Batchelor et al. 2010). Sorensen et al. (2009) evaluated after a single dose of cediranib the correlation between clinical outcome in glioblastoma patients and MRI changes in vascular permeability/flow and in microvessel outcome, and the change of circulating collagen IV levels, They demonstrated that the changes in these parameters and their combination in a "vascular normalization index" correlate with duration of overall survival and/or progression-free survival.

The inhibitors of VEGF in the therapy of central nervous system malignancy normalizes tumor vasculature and decrease tumor interstitial pressure, leading to an improved access of cytoreductive drugs and radiotherapy efficacy, due to an increased oxygen delivery (Mc Gee et al. 2010). However, these agents may also restore the low permeability characteristics of normal brain microvasculature, counteracting beneficial effects. The presence of an intact blood-brain barrier in some areas of the tumor and the presence of a partially functional blood-brain barrier in other areas of the tumor can prevent the delivery of therapeutic compounds.

VEGFR-2 blockade can lead to the upregualtion of Ang-1 that increase pericyte coverage of the vessels (Winkler et al. 2004). Ang-2 plays a more important role in tumor angiogenesis than it does in normal angiogenesis. As an antagonist for Ang-1, it is responsible for blood vessel destabilization in vasculature surrounding tumors.

In glioblastoma patients, the Ang-1/Ang-2 ratio correlates with survival (Sie et al. 2009) and vascular normalization, whereas high Ang-2 levels correlate with resistance to anti-VEGF therapy (Batchelor et al. 2010). Moreover, blockade of VEGF signaling with the VEGFR tyrosine kinase inhibitor cediranib significantly reduced levels of Ang-2 in some patients, even if the decrease was transient and modest (Batchelor et al. 2010). Chae et al. (2010) expressed Ang-2 in an orthotopic glioma model and demonstrated that ectopic expression of Ang-2 had no effect on vascular permeability, tumor growth, or survival, but it resulted in higher vascular density, with dilated vessels and reduced mural cell coverage. When combined with anti-VEGFR-2 treatment, Ang-2 destabilized vessels and compromised the survival benefit of VEGFR-2 inhibition by increasing vascular permeability, suggesting that VEGFR-2 inhibition normalized tumor vasculature, whereas ectopic expression of Ang-2 diminished the beneficial effects of VEGFR-2 blockade by inhibiting vessel normalization. In transplantable tumor models, blockade of stromal Ang-2 induces tightening of endothelial cell junctions, pericyte coverage, and remodeling to smaller vessels, reducing tumor growth (Falcon et al. 2009; Nasarre et al. 2009).

PDGF over-expression can normalize tumor vessels and increase drug delivery (Liu et al. 2011), while PDGFR blockade improves drug delivery and chemotherapy (Helberg et al. 2010). Specific silencing or blocking of nNOS in tumor cells restores a NO gradient such that NO is concentrated primarily around blood vessels, resulting in a more normal vessel phenotype (Kashiwagi et al. 2008). PlGF blockade normalizes tumor vessels by shifting the polarization of TAMs from a M2-like phenotype to an M1-like phenotype (Rolny et al. 2011). Preclinical studies using models of spontaneous hepatocellular carcinoma have shown that PlGF blockade induces vessel normalization (Van de Veire et al., 2010). Inhibition of regulator G-protein signaling 5, produced by activated pericytes and hypoxic endothelial cells, results in vessel normalization via vessel maturation (Hamzah et al. 2008).

4.6 Drawbacks of Anti-angiogenesis Therapies

Although VEGF and VEGFR are validated targets in solid tumors, the eventual resistance of tumors to these agents, through intrinsic and acquired mechanisms (Ellis and Hicklin 2008), and the subsequent progression of disease continues to be a clinically significant problem (Kerbel et al. 2001; Miller et al. 2003, 2005). Preclinical studies have demonstrated that intrinsic resistance may result from an absence of VEGF or VEGFR in tumors from certain organ sites (Karashima et al. 2007). Alternatively, a tumor may exhibit acquired resistance by co-opting existing blood vessels from vasculature-rich organs, such as the lungs, liver or brain (Leenders et al. 2004). The development of hypoxia-resistant tumor subpopulations (through mutation) which can outgrow the sensitive tumor cells (Yu et al. 2002) or the selection of more mature, stable vessels that are intrinsically less responsive to anti-angiogenic treatment (Glade Bender et al. 2004) could also cause resistance. Tumor -associated endothelial cells may also be a source of resistance due to various cytogenetic abnormalities (Hida et al. 2004).

A reduced response to VEGF-targeted therapies may also originate from stroma compartment. The stroma contains a heterogeneous population of cells, including fibroblasts and BMDC (Dong et al. 2004; Shojaei and Ferrara 2007). BMDC have been shown to support tumor growth by a variety of mechanisms, including by direct contribution to the tumor vasculature (Santerelli et al. 2006) and by releasing pro-angiogenic factors, including VEGF (Liang et al. 2006). Moreover, myeloid cells are recruited by resistant tumors and contribute in VEGF-independent angiogenesis via release of angiogenic factors, such as Bv8 (Shojaei et al. 2007). G-CSF is a key regulator of Bv8 expression and inhibition of G-CSF can suppress tumor growth as single agent or in combination with anti-VEGF antibody to similar extent that was observed with anti-Bv8 treatment (Shojaei et al. 2009). Finally, myeloid cells promote tumor growth via suppression of immune response (Huang et al. 2006).

Tumor-associated fibroblasts have been implicated in angiogenesis and resistance due to their production of SDF-1, a molecule that promotes the recruitment of endothelial progenitor cells and myeloid cells into the tumor microenvironment (Shojaei and Ferrara 2007). Moreover, tumor-associated fibroblasts are a source of other angiogenic cytokines, such as VEGF, CXCL12, and PDGF (Crawford et al. 2009).

When resistance to VEGF becomes apparent and tumor growth is observed despite inhibition of the VEGF pathway it is postulated that the tumor, under conditions such as hypoxia, utilizes compensatory mechanisms involving alternative, previously 're-dundant' pro-angiogenic pathways and factors (Ellis and Hicklin 2008; Casanovas et al. 2005). Indeed, it has been demonstrated that after an initial response to an anti-VEGFR-2 monoclonal antibody (cediranib) and subsequent angiogenic rebound associated with increased FGF expression, use of an FGF-trap minimized the acquired resistance to the VEGF-targeted therapy (Ellis and Hicklin 2008; Casanovas et al. 2005; Batchelor et al. 2007). This suggests that the FGF pathway was triggered by the tumor as a compensatory mechanism to elicit continued angiogenesis. It has also been shown that the number of angiogenic molecules expressed in a tumor increases with malignant progression, resulting in a decreased dependence on VEGF in certain advanced tumors (Carmeliet 2005).

While it has always been assumed that tumors produce multiple pro-angiogenic factors, it was also assumed that due to angiogenic redundancy production of certain factors would only be switched on as needed, for example, upon challenge with a certain TKI, allowing angiogenesis to continue (Arbiser 2007). However, it has now been shown that multiple TK may in fact play a normal part in tumor angiogenesis, even when not challenged by a TKI (Nissen et al. 2007; Arbiser 2007). Preclinical studies have shown synergy between PDGF-BB and FGF-2 in the stimulation of angiogenesis, with rapid regression of neo-vascularization observed when either FGF-2 or PDGF-BB was present alone (Nissen et al. 2007). FGF-2 has also been shown to up-regulate the expression of PDGF-BB in epithelial cells and the production of certain enzymes required for tumorigenesis (Nissen et al. 2007).

Given the number of receptors, signaling pathways and molecules involved in angiogenesis, there is potential extensive redundancy among proangiogenic factors and crosstalk occurs within various pro-angiogenic and anti-angiogenic signaling pathways (Relf et al. 1997). The concept of angiogenic redundancy calls into question the effectiveness of targeted therapies that target only one pro-angiogenic factor.

Anti-angiogenic targeting of VEGF signaling induces, after a period of transitory response, a re-vascularization and increased invasiveness of the tumor (Casanovas et al. 2005). Magnetic resonance imaging in a substet of glioblastoma multiforme patients has documented the development of multifocal recurrence of tumors during the course of therapy with bevacizumab (Norden et al. 2008).

Two papers reported that antiangiogenic therapy promotes tumor invasion and metastasis (Paez-Ribes et al. 2009; Ebos et al. 2009). Sunitinib, a multi-targeted RTK that potently inhibits VEGF and PDGF signaling, and the anti-VEGFR-2 antibody DC101 stimulated the invasive behavior of tumor cells despite their inhibition of primary tumor growth and increased overall survival in some cases (Paez-Ribes et al. 2009; Ebos et al. 2009)

Ebos et al. (2009) injected human metastatic breast cancer 231/LM2–4^{LUC+} cells expressing luciferase into tail vein of severe combined immunodeficiency (SCID) mice and assessed tumor burden by bioluminescence. They observed that mice receiving sunitinib for 7 days either before or after injection of tumor cells showed an accelerated tumor growth resulting in a shorter survival. The same was observed in *nu/nu* mice treated with sorafenib, a multikinase inhibitor that target both Raf and VEGF and PDGF RTK signaling (Wilhelm et al. 2008), or SU10944 and with human MeWo melanoma cells. Moreover, mice treated for 7 days just after resection of an orthotopic implanted tumor showed an accelerated tumor growth, shorter survival, and the increased overall tumor burden corresponded to a diffuse metastatic process in multiple organs.

Paez-Ribes et al. (2009), by using the RIP 1-TAG 2 model of pancreatic cancer, demonstrated that the tumor front of invasion is more frequently intermingled with the surrounding tissue without encapsulation already after 7 days of treatment with DC 101 and this effect was accentuated by longer therapy and persisted after cessation of therapy. Moreover, tumor carrying a tumor cell-specific deletion of the VEGF-A gene showed increased invasiveness and distant metastasis. Finally, treatment with sunitinib produced significant survival benefit but induced also a more aggressive phenotype characterized by widespread tumor infiltration and hematogenic metastasis associated with hypoxia in the primary tumor (Paez-Ribes et al. 2009).

Increased invasiveness might result from enhanced expression of various angiogenic cytokines induced by the treatment, such as VEGF and PlGF, or recruitment of endothelial progenitor cells that promote the formation of a pre-metastatic niche (Ebos et al. 2009). Hypoxia-driven effects may be also involved, because hypoxia generated by angiogenesis inhibition triggers pathways that make tumors more aggressive and metastatic and less sensitive to anti-angiogenic treatment (Paez-Ribes et al. 2009; Ebos et al. 2009). For example, therapy-induced hypoxia increase in tumor hypoxia and HIF-1 α expression following VEGF inhibition can lead to increased c-met expression (Pennacchietti et al., 2003), increased IL-6 expression (Saidi et al. 2009) and activation and/or up-regulation of MMP (Cairns et al. 2003).

Sunitinib may disrupt vascular integrity through pericyte detachment mediated by inhibition of PDGFR-β, thus facilitating intravasation and extravasation of tumor cells. To support this hypothesis, Xian et al. (2006) demonstrated that vessels with poor pericyte coverage favor metastatic process.

The incidence and the number of tumor micrometastasis was increased in liver specimens of sunitinib-treated *vs* control mice (Hu-Lowe et al. 2008), while no increase in lymphatic metastasis was observed after sunitinib treatment, in contrast to the anti-VEGFR-2 antibody. This difference could be due to the fact that sunitinib blocks not only VEGFR-2 and PDGFR, but also lymphatic vessel-related VEGFR-3 (Roskoski 2007).

The findings of Paez-Ribes et al. (2009) and Ebos et al. (2009) may explain why in clinical experience VEGF-targeted therapy is followed by a transitory period of primary tumor growth inhibition and prolongation of progression-free survival, tumor relapse as more invasive metastatic disease. For example, in human glioblastoma multiforme patients treated with bevacizumab in combination with chemotherapy, it has been described tumor relapse and/or regrowth accompained by a high rate of diffuse infiltrative lesions (Narayana et al. 2008). Kunkel et al. (2001) have previously demonstrated that treatment of mice with a monoclonal antibody against VEGFR-2 induced a shift in glioblastoma tumor phenotype toward enhanced migration and metastasis.

Clinical anti-metastatic activity of sunitinib has been reported for renal cell carcinoma in different metastatic sites, i.e. pancreas, heart, brain and prostate (Medioni et al. 2009; Szmitz et al. 2009; Fokt et al. 2009; Kooutras et al. 2007). Experimental evidence is provided for the beneficial effect of combined sunitinib and radiotherapy for an orthotopic murine model of breast cancer metastasis in bone treatment (Zwolak et al. 2008). Moreover, sunitinib monotherapy is reported to efficiently inhibit tumor growth and osteolysis in another breast cancer bone metastasis model (Murray et al. 2003).

In the last 35 years, it has been estimated that >200 companies have worked and are still working in the area of angiogenesis and several of the compounds that modulate angiogenesis are currently being evaluated in clinical trials. Even if the majority of pre-clinical studies have shown that the growth of all experimental tumors can be effectively inhibited by various anti-angiogenic agents, the clinical benefits of anti-angiogenic treatments are relatively modest, and in the majority of cases, the drugs merely slow down tumor progression and prolong survival by only a few more months.

Clinical studies have shown benefits in relapsed-free survival for metastatic colorectal cancer, advanced non-small cell lung cancer, renal cell carcinoma, hepatocellular carcinoma, metastatic breast cancer, GIST and in glioblastoma (Ellis and Hicklin 2008; Shojaei and Ferrara 2008), but overall survival benefit has not yet been seen (Miller 2003), with the exception of bevacizumab treatment in renal cell carcinoma as a single agent (Yang et al. 2003), or in metastatic breast cancer in combination with a taxane chemotherapy (Miller et al. 2007). The most impressive clinical response occurred in the low dose bevacizumab plus chemotherapy with a statistically significant median overall survival (21.5 months) versus fluorouracil/leucovorin alone (13.9 months) or high-dose bevacizumab plus fluorouracil/leuocovirin (16.1 months) (Kabbinavar et al. 2003).

Autocrine VEGF signaling to promote malignant cell survival is also a common feature in haematological malignancies, suggesting that anti-VEGF/VEGFR targeted therapy would promote direct killing of tumor cells, as well as inhibit angiogenesis. VEGF-directed therapy has been investigated also in hematological malignancies, most commonly in acute myeloid leukemia, myelodysplastic syndrome, and in non-Hodgkin lymphoma. Clinical trials involving anti-VEGF agents induce only a modest improvement in overall survival, measurable in weeks to just a few months, and various tumors respond differently in human patients to these agents.

In summary, different synergistic causes are involved in the establishment of inefficacy of anti-angiogenic treatment:

(i) Lack of understanding of which patients will show the benefit of these agents and occurrence of drug resistance (Miller 2003; Jain et al. 2006; Bergers and Hanahan 2008). This is due to the absence of reliable surrogate markers of angiogenesis and anti-angiogenesis to demonstrate the efficacy of anti-angiogenic agents in clinical trials and for the monitoring of these agents (Ribatti, 2010).

(ii) Endothelial cells isolated from various tumors acquired genotype alterations, exhibiting aneuploidy, abnormal multiple chromosomes, and aberrant chromosomal architecture (Hida et al. 2004). Proximity of tumor cells and endothelial cells within the tumor microenvironment may be responsible for the genotype alterations (Hida and Klagsbrun 2005). Genetic alteration of endothelial cells leads to altered anti-angiogenic targets and resistance.

(iii) Anti-angiogenic therapies may sometimes promote invasion and metastasis (Ribatti 2011). Sunitinib, a multi-targeted receptor tyrosine kinase inhibitor of VEGF and PDGF signaling and the anti-VEGFR-2 antibody DC101 stimulate the invasive behavior of tumor cells despite their inhibition of primary tumor growth and increased overall survival in some cases (Paez-Ribes et al. 2009; Ebos et al. 2009).

(iv) Inherent or acquired resistance to anti-VEGF drugs can occur in patients, leading in some cases to a lack of response and in others to disease recurrence, although discontinuation of the therapy at the time of progression is a factor limiting the effectiveness of anti-angiogenic therapy. In the meantime, prolonged VEGF leads to vascular pruning and endothelial cell apoptosis, release of cytokines by host cells, which may promote tumor re-growth.

(v) In most tumors, the vasculature is altered showing increased permeability, vessel dilatation, decreased/abnormal pericyte coverage and abnormal basement membrane structure. While VEGF neutralization can initially limit tumor proliferation due to its anti-angiogenic effect, it can also result in transient vascular normalization with improved oxygenation and perfusion (Jain 2005), favoring drug delivery. However, in gliomas normalization of the vascular bed involves restoration of the blood-brain barrier, thereby hampering, instead of enhancing, the delivery of therapeutic compounds to tumor cells (Claes et al. 2008).

(vi) Prolonged VEGF inhibition increases local hypoxia leading to systemic secretion of other angiogenic cytokines, such as FGF-2 and SDF-1α, which may promote cancer re-growth and metastasis (Casanovas et al. 2005). An analysis of human breast cancer biopsies revealed that late-stage breast cancers expressed several angiogenic cytokines in contrast to earlier stage lesions, which preferentially expressed VEGF (Relf et al. 1997). Up-regulation of VEGF and PlGF has been reported in patients treated with antiangiogenic agents targeting the VEGF pathway and FGF-2 has been found to be increased in the blood of patients relapsed from treatment using VEGFR inhibitors (Batchelor et al. 2007).

References

Abdel-Majid RM, Marshall JS (2004) Prostaglandin E2 induces degranulation-independent production of vascular endothelial growth factor by human mast cells. J Immunol 172:1227–1236

Abdollahi A, Lipson KE, Han X et al (2003) SU5416 and SU6668 attenuate the angiogenic effects of radiation-induced tumor cell growth factor production and amplify the direct anti-endothelial action of radiation in vitro. Cancer Res 63:3755–3763

Abdollahi A, Hahnfeldt P, Maercker C et al (2004) Endostatin's antiangiogenic signaling network. Mol Cell 13:649–663

Abramsson A, Lindblom P, Betsholtz C et al (2003) Endothelial and nonendothelial sources of PDGF-B regulate pericyte recruitment and influence vascular pattern formation in tumors. J Clin Invest 112:1142–1151

Achen MG, Stacker SA (2008) Molecular control of lymphatic metastasis. Ann N Y Acad Sci 1131:225–234

Adachi Y, Nakamura H, Kitamura Y et al (2007) Lymphatic vessel density in pulmonary adenocarcinoma immunohistochemically evaluated with anti-podoplanin or anti-D2-40 antibody is correlated with lymphatic invasion or lymph node metastases. Pathol Intern 57:171–177

Adjei AA, Molina JR, Mandrekar SJ et al (2007) Phase I trial of sorafenib in combination with gefitinib in patients with refractory or recurrent non-small cell lung cancer. Clin Cancer Res 13:2684–2691

Ahmad SA, Liu W, Jung YD et al (2001) The effects of angiopoietin-1 and -2 on tumor growth and angiogenesis in human colon cancer. Cancer Res 61:1255–1259

Aicher A, Heeschen C, Midner-Rihm C et al (2003) Essential role of endothelial nitric oxide synthase for mobilization of stem and progenitor cells. Nat Med 9:1370–1376

Al-Nedawi K, Meehan B, Micallef J et al (2008) Intercellular transfer of the oncogenic receptor EGFRvII by microvesicles derived from tumour cells. Nat Cell Biol 10:619–624

Al-Nedawi K, Meehan B, Kerbel RS et al (2009) Endothelial expression of autocrine VEGF upon the uptake of tumor-derived micorvesicles containing oncogenic EGFR. Proc Natl Acad Sci U S A 106:3794–3799

Alitalo K, Tammela T, Petrova TV (2005) Lymphangiogenesis in development and human disease. Nature Insights 438:946–953

Allen TM (2002) Ligand-targeted therapeutics in anticancer therapy. Nat Rev Cancer 2:750–763

Allen TM, Cheng WW, Hare JI et al (2006) Pharmacokinetics and pharmacodynamics of lipidic nano-particles in cancer. Anticancer Agents Med Chem 6:513–523

Ancelin M, Chollet-Martin S, Herve MA et al (2004) Vascular endothelial growth factor VEGF189 induces human neutrophil chemotaxis in extravascular tissue via an autocrine amplification mechanism. Lab Invest 8:502–512

Anghelina M, Moldovan L, Zabuawala T et al (2006) A subpopulation of peritoneal macrophages form capillary-like lumens and branching patterns in vitro. J Cell Mol Med 10:708–715

Angiolillo AL, Sgadari C, Taub DD et al (1995) Human interferon-inducible protein 10 is a potent inhibitor of angiogenesis in vivo. J Exp Med 182:155–162

D. Ribatti, *Morphofunctional Aspects of Tumor Microcirculation*,
DOI 10.1007/978-94-007-4936-8, © Springer Science+Business Media Dordrecht 2012

Annunziata CM, Walker AJ, Minasian L et al (2010) Vandetanib, designed to inhibit VEGFR2 and EGFR signaling, had no clinical activity as monotherapy for recurrent ovarian cancer and no detectable modulation of VEGFR2. Clin Cancer Res 16:664–672

Aoki M, Pawankar R, Niimi Y et al (2003) Mast cells in basal cell carcinoma express VEGF, IL-8 and RANTES. Int Arch Allergy Immunol 130:216–223

Arbiser J (2007) Why targeted therapy hasn't worked in advanced cancer. J Clin Invest 117:2762–2765

Arigami T, Natsugoe S, Uenosono Y et al (2009) Vascular endothelial growth factor-C and -D expression correlates with lymph node micrometastasis in pNo early gastric cancer. J Surg Oncol 99:148–153

Asahara T, Murohara T, Sullivan A et al (1997) Isolation of putative progenitor endothelial cells for angiogenesis. Science 275:964–947

Asahara T, Masuda H, Takahashi T et al (1999a) Bone marrow origin of endothelial progenitor cells responsible for postnatal vasculogenesis in physiological and pathological neovascularization. Circ Res 85:221–228

Asahara T, Takahashi T, Masuda H et al (1999b) VEGF contributes to postnatal neovascularization by mobilizing bone-marrow derived endothelial progenitor cells. EMBO J 18:3964–3972

Au P, Tam J, Fukumura D et al (2008) Bone marrow-derived mesenchymal stem cells facilitate engineering of long-lasting functional vasculature. Blood 111:4551–4558

Aurora AB, Baluk P, Zhang D et al (2005) Immune complex-dependent remodeling of the airway vasculature in response to a chronic bacterial infection. J Immunol 175:6319–6326

Aznavoorian S, Murphy AN, Stetler-Stevenson WG et al (1993) Molecular aspects of tumor cell invasion and metastasis. Cancer 71:1638–1683

Baldewijns MM, Roskams T, Ballet V et al (2009) A low frequency of lymph node metastasis in clear-cell renal cell carcinoma is related to low lymphangiogenic activity. BJU Int. doi:10.1111/j.1464–410X.2008.08272.x

Balkwill F, Mantovani A (2001) Inflammation and cancer: back to Virchow? Lancet 357:539–545

Baluk P, Morikawa S, Haskell A et al (2003) Abnormalities of basement membrane on blood vessels and endothelial sprouts in tumors. Am J Pathol 163:1801–1815

Baluk P, Hashizume H, Mc Donald DM (2005) Cellular abnormalities of blood vessels as targets in cancer. Curr Opin Genet Dev 15:102–111

Bancherau J, Steinman RM (1998) Dendritic cells and the control of immunity. Nature 392:245–252

Banchereau J, Briere F, Caux C et al (2000) Immunobiology of dendritic cells. Ann Rev Immunol 18:767–811

Bardin N, George F, Mutin M et al (1996) S-endo 1, a pan-endothelial monoclonal antibody recognizing a novel human endothelial antigen. Tissue Antigens 48:531–539

Barleon B, Sozzani S, Zhou D et al (1996) Migration of human monocytes in response to vascular endothelial growth factor/(VEGF) is mediated via the VEGF receptor flt-1. Blood 87:3336–3343

Barrera P, Blom A, van Lent PL et al (2000) Synovial macrophage depletion with clodronate-containing liposomes in rheumatoid arthritis. Arthritis Rheum 43:1951–1959

Barlett JB, Dredge K, Dalgleish AG (2004) The evolution of thalidomide and its IMiD derivatives as anticancer agents. Nat Rev Cancer 4:314–322

Bartlett MR, Underwood PA, Parish CR (1995) Comparative analysis of the ability of leucocytes, endothelial cells and platelets to degrade the subendothelial basement membrane: evidence for cytokine dependence and detection of a novel sulfatase. Immunol Cell Biol 73:113–124

Batchelor TT, Sorensen AG, di Tomaso E et al (2007) AZD2171, a pan-VEGF receptor tyrosine kinase inhibitor, normalizes tumor vasculature and alleviates edema in glioblastoma patients. Cancer Cell 11:83–95

Batchelor TT, Duda DG, di Tomaso E et al (2010) Phase II study of cediranib, an oral pan-vascular endothelial growth factor receptor tyrosine kinase inhibitor, in patients with recurrent glioblastoma. J Clin Oncol 28:2817–2823

Bazzoni F, Cassatella MA, Rossi F et al (1991) Phagocytosing neutrophils produce and release high amounts of the neutrophil-activating peptide 1/interleukin 8. J Exp Med 173:771–774

Beasley NJP, Prevo R, Banerji S et al (2002) Intratumoral lymphangiogenesis and metastasis in head and neck cancer. Cancer Res 62:1315–1320

Beckermann BM, Kallifatidis G, Froth A et al (2008) VEGF expression by mesenchymal stem cells contributes to angiogenesis in pancreatic carcinoma. Br J Cancer 99:622–631

Beerepoot LV, Mehra N, Bermaat JS et al (2004) Increased levels of viable circulating endothelial cells are an indicator of prgressive disease in cancer patients. Ann Oncol 15:139–145

Bellmunt J (2009) Future developments in renal cell carcinoma. Ann Oncol 20(Suppl 1):i13–i17

Benelli R, Morini M, Carrozzino F et al (2002) Neutrophils as a key cellular target for angiostatin: implications for regulation of angiogenesis and inflammation. FASEB J 16:267–269

Benitez-Bribiesca L, Wong A, Utrera D et al (2001) The role of mast cell tryptase in neoangiogenesis of premalignant and malignant lesions of the uterine cervix. J Histochem Cytochem 49:1061–1062

Bergers G, Hanahan D (2008) Modes of resistance on antiangiogenic therapy. Nat Rev Cancer 8:592–603

Bergers G, Song S, Mayer-Morse N et al (2003) Benefits of targeting both pericytes and endothelial cells in the tumor vasculature with kinase inhibitors. J Clin Invest 111:1287–1295

Bertolini F, Shaked Y, Mancuso P et al (2006) The multifaceted circulating endothelial cell in cancer: towards marker and target identification. Nat Rev Cancer 6:835–845

Bingle L, Brown NJ, Lewis CE (2002) The role of tumor associated macrophages in tumor progression; implications for new anticancer therapies. J Pathol 196:254–265

Black WC, Welch HG (1993) Advances in diagnostic imaging and overestimations of disease prevalence and the benefits of therapy. N Engl J Med 328:1237–1243

Blair RJ, Meng H, Marchese MJ et al (1997) Human mast cells stimulate vascular tube formation: tryptase is a novel potent angiogenic factor. J Clin Invest 99:2691–2700

Blinder V, Fischer SG (2008) The role of environmental factors in the etiology of lymphoma. Cancer Invest 26:306–316

Bocci G, Francia G, Man S et al (2003) Thrombospondin-1, a mediator of the antiangiogenic effects of low-dose metronomic chemotherapy. Proc Natl Acad Sci U S A 100:12917–12922

Boehm T, Folkman J, Browder T et al (1997) Antiangiogenic therapy of experimental cancer does not induce acquired drug resistance. Nature 390:404–407

Boesiger J, Tsai M, Maurer M et al (1998) Mast cells can secrete vascular permeability factor/vascular endothelial cell growth factor and exhibit enhanced release after immunoglobulin E-dependent upregulation of fc epsilon receptor I expression. J Exp Med 188:1135–1145

Bonsib SM (2006) Renal lymphatics, and lymphatic involvement in sinus vein invasive (pT3b) clear cell renal cell carcinoma: a study of 40 cases. Mod Pathol 19:746–753

Bourbié-Vaudanie S, Blanchard N, Hivroz C et al (2006) Dendritic cells can turn CD4+ T lymphocytes into vascular endothelial growth factor-carrying cells by intercellular neuropilin-1 transfer. J Immunol 177:1460–1469

Bornstein P (2001) Thrombospondins as matricellular modulators of cell function. J Clin Invest 107:929–934

Braun M, Flucke U, Debald M et al (2008) Detection of lymphovascular invasion in early breast cancer by D2-40 (podoplanin): a clinically useful predictor for axillary lymph node metastases. Breast Cancer Res Treat 112:503–511

Brouty-Boye D, Zetter BR (1980) Inhibition of cell motility by interferon. Science 208:16–18

Browder T, Butterfield CE, Kraling BM et al (2000) Antiangiogenic scheduling of chemotherapy improves efficacy against experimental drug-resistant cancer. Cancer Res 60:1878–1886

Bowrey PF, King J, Magarey C et al (2000) Histamine, mast cells and tumour cell proliferation in breast cancer: does preoperative cimetidine administration have an effect? Br J Cancer 82:167–170

Buckstein R, Kerbel RS, Shaked Y et al (2006) High-dose celecoxib and metronomic "low-dose" cyclophosphamide is an effective and safe therapy in patients with relapsed and refractory aggressive histology non-Hodgkin's lymphoma. Clin Cancer Res 12:5190–5918

Bukowski R, Eisen T, Szczylik C et al (2007) Final results of the randomzied Phase III trial of sorefenib in advanced renal cell carcinoma: survival and biomarker analysis. J Clin Oncol 25(June 20 Suppl):5023 (Abstract)

Burger RA, Brady MF, Bookman MA et al (2010) Phase III trial of bevacizumab (BEV) in the primary treatment of advanced epithelial ovarian cancer (EOC), primary peritoneal cancer (PPC), or fallopian tube cancer (FTC): a gynecologic oncology group study. J Clin Oncol 28 (Abstract LBA1)

Burri PH, Djonov V (2002) Intussusceptive angiogenesis, the alternative to capillary spruting. Mol Aspects Med 23:S1–S27

Bussolati B, Deambrosis I, Russo S et al (2003) Altered angiogenesis and survival in human tumor-derived endothelial cells. FASEB J 17:1159–1161

Bussolati B, Grange C, Bruno S et al (2006) Neural-cell adhesion molecule (NCAM) expression by immature and tumor-derived endothelial cells favors cell organization into capillary-like structures. Exp Cell Res 312:913–924

Cairns RA, Khokha R, Hill RP (2003) Molecular mechanisms of tumor growth and metastasis: an integrated view. Curr Mol Med 3:659–671

Calcano G, Lee J, Kikly K et al (1994) Neutrophil and B cell expansion in mice that lack murine IL-8 receptor homolog. Science 265:682–684

Cao Y (2005) Emerging mechanisms of tumor lymphangiogenesis and lymphatic metastasis. Nature Rev Cancer 7:192–198

Cao Y (2009) Tumor angiogenesis and molecular targets for therapy. Front Biosci 14:3962–3973

Cardone RA, Casavola V, Reshkin SJ (2005) The role of disturbed pH dynamics and the Na^+/H^+ exchanger in metastasis. Nature Rev Cancer 5:786–795

Carmeliet P (2005) Angiogenesis in life, disease and medicine. Nature 438:932–936

Carmeliet P, Dor Y, Herbert JM, Fukumura D et al (1998) Role of HIF-1 in hypoxia-mediated apoptosis, cell proliferation and tumor angiogenesis. Nature 394:485–490

Casanovas O, Hicklin DJ, Bergers G et al (2005) Drug resistance by evasion of antiangiogenic targeting of VEGF signaling in late-stage pancreatic islet tumors. Cancer Cell 8:299–309

Case J, Mead LE, Bessler WK et al (2007) Human $CD34^+$ $AC133^+$ $VEGFR2^+$ cells are not endothelial progenitor cells but distinct, primitive hematopoietic progenitors. Exp Hematol 35:1109–1118

Cavallo F, Quaglino E, Cifaldi L et al (2001) Interleukin 12-activated lymphocytes influence tumor genetic programs. Cancer Res 61:3518–3523

Chae SS, Paik JH, Furneaux H et al (2004) Requirement for sphingosine 1-phosphate receptor-1 in tumor angiogenesis demonstrated by in vivo RNA interference. J Clin Invest 114:1082–1089

Chae SS, Kamoun WS, Farrar CT et al (2010) Angiopoietin-2 interferes with anti-VEGFR2-induced vessel normalization and survival benefit in mice bearing gliomas. Clin Cancer Res 16:3618–3627

Chandrasekaean L, He CZ, Al-Barazi H et al (2000) Cell contact-dependent activation of alpha3beta1 integrin modulates endothelial cell responses to thrombospondin-1. Mol Biol Cell 11:2855–2900

Chang YS, di Tomaso E, Mc Donald DM et al (2000) Mosaic blood vessels in tumors: frequency of cancer cells in contact with flowing blood. Proc Natl Acad Sci U S A 97:14608–14613

Chantrain CF, Shimada H, Jodele S et al (2004) Stromal matrix metalloproteinase-9 regulates the vascular architecture in neuroblastoma by promoting pericyte recruitment. Cancer Res 64:1675–1686

Chavakis T, Cines DB, Rhee JS et al (2004) Regulation of neovascularization by human neutrophil peptides (alpha-defensins): a link between inflammation and angiogenesis. FASEB J 18:1306–1308

Chen L, Tredget EE, Wu PYG et al (2008) Paracrine factors of mesenchymal stem cells recruit macrophages and endothelial lineage cells and enhance wound healing. PloS One 3:e1886

Chen W, Chen M, Liao Z et al (2009) Lymphatic vessel density as predictive marker for the local recurrence of rectal cancer. Dis Colon Rectum 52:513–519

Chen ZY, Wei W, Guo ZX et al (2011) Morphologic classification of microvessels in hepatocellular carcinoma is associated with a prognosis after resection. J Gastroenterol Hepatol 26:91–101

Childs S, Chen JN, Garrity DM et al (2002) Patterning angiogenesis in the zebrafish embryo. Development 129:973–982

Choi WWL, Lewis MM, Lawson D et al (2005) Angiogenic and lymphangiogenic microvessel density in breast carcinoma: correlation with clinicopathologic parameters and VEGF-family gene expression. Mod Pathol 18:143–152

Chuang WY, Yeh CJ, Wu YC et al (2009) Tumor cell expression of podoplanin correlates with nodal metastasis in esophageal squamous cell carcinoma. Histol Histopathol 24:1021–1027

Claes A, Wesseling P, Jeuken J et al (2008) Antiangiogenic compounds interfere with chemotherapy of brain tumors due to vessel normalization. Mol Cancer Ther 7:71–78

Claffey KP, Brown LF, Del Aguila LF et al (1996) Expression of vascular permeability factor/vascular endothelial growth factor by melanoma cells increases tumor growth, angiogenesis, and experimental metastasis. Cancer Res 56:172–181

Colombo G, Curnis F, De Mori GM et al (2002) Structure-activity relationships of linear and cyclic peptides containing the NGR tumor-homing motif. J Biol Chem 277:47891–4797

Colombo MP, Trinchieri G (2002) Interleukin-12 in anti-tumor immunity and immunotherapy. Cytokine Growth Factor Rev 13:155–168

Colorado PC, Torre A, Kamphaus G et al (2000) Anti-angiogenic cues from vascular basement membrane collagen. Cancer Res 60:2520–2526

Cook N, Basu B, Biswas S et al (2010) A phase 2 study of vatalanib in metastatic melanoma patients. Eur J Cancer 46:2671–2673

Coussens LM, Werb Z (1996) Matrix metalloproteinases and the development of cancer. Chem Biol 3:895–904

Crawford Y, Kasman I, Yu L et al (2009) PDGF-C mediates the angiogenic and tumorigenic properties of fibroblasts associated with tumors refractory to anti-VEGF treatment. Cancer Cell 15:21–34

Crivellato E, Nico B, Vacca A et al (2003) B-cell non-Hodgkin's lymphomas express heterogeneous patterns of neovascularization. Haematologica 88:671–678

Curnis F, Arrigoni G, Sacchi A et al (2002) Differential binding of drugs containing the NGR motif to CD13 isoforms in tumor vessels, epithelia, and myeloid cells. Cancer Res 62:867–874

D'Amato RJ, Loughman MS, Flynn E et al (1994) Thalidomide is an inhibitor of angiogenesis. Proc Natl Acad Sci U S A 91:4082–4085

Dadras SS, Lange-Asschenfeldt B, Velasco P et al (2005) Tumor lymphangiogenesis predicts melanoma metastasis to sentinel lymph nodes. Mod Pathol 18:1232–1242

Dal Lago L, D'Hondt V, Awada A (2008) Selected combination therapy with sorafenib: a review of clinical data and perspectives in advanced solid tumors. Oncologist 13:845–858

Daldrup H, Shames DM, Wendland M et al (1998) Correlation of dynamic contrast-enhanced MR imaging with histologic tumor grade: comparison of macromolecular and small-molecular contrast media. Am J Roentgenol 171:941–949

Damert A, Machein M, Breier G et al (1997) Up-regulation of vascular endothelial growth factor expression in a rat glioma is conferred by two distinct hypoxia-driven mechanisms. Cancer Res 57:3860–3864

de Boer R, Humblet Y, Wolf J et al (2009) An open-label study of vandetanib with pemetrexed in patients with previously treated non-small-cell lung cancer. Ann Oncol 20:486–491

Della Porta MG, Malcovati L, Rigolin GM et al (2008) Immunophenotypic, cytogenetic and functional characterization of circulating endothelial cells in myelodysplastic syndromes. Leukemia 22:530–537

De Luisi A, Ferrucci A, Coluccia AML et al (2011) Lenalidomide restrains motility and overangiogenic potential of bone marrow endothelial cells in patients with active multiple myeloma. Clin Cancer Res 17:1935–1946

De Palma M, Venneri MA, Roca C et al (2003) Targeting exogenous genes to tumor angiogenesis by transplantation of genetically modified hematopoietic stem cells. Nat Med 9:789–795

De Palma M, Venneri MA, Galli R et al (2005) Tie 2 identifies a hematopoietic lineage of proangiogenic monocytes required for tumor vessel formation and a mesenchymal population of pericyte progenitors. Cancer Cell 8:211–226

De Palma M, Mazzieri R, Politi LS et al (2008) Tumor-targeted interferon-α delivery by Tie2-expressing monocytes inhibits tumor growth and metastasis. Cancer Cell 14:299–311

de Paulis A, Prevete N, Rossi FW et al (2006) Expression and functions of vascular endothelial growth factors and their receptors in human basophils. J Immunol 177:7322–7331

Demetri GD, van Oosterom AT, Garrett CR et al (2006) Efficacy and safety of sunitinib in patients with advanced gastrointestinal stromal tumour after failure of imatinib: a randomized controlled trial. Lancet 368:1329–1338

Denekamp J (1982) Endothelial cell proliferation as a novel approach to targeting tumor therapy. Br J Cancer 45:136–139

Denhardt DT, Noda M, O'Regan AW et al (2001) Osteopontin as a means to cope with environmental insults: regulation of inflammation, tissue remodeling, and cell survival. J Clin Invest 107:1055–1061

Detmar M, Brown LF, Schon MP et al (1998) Increased microvascular density and enhanced leukocyte rolling and adhesion in the skin of VEGF transgenic mice. J Invest Dermatol 111:1–6

Detoraki A, Staiano RI, Granata F et al (2009) Vascular endothelial growth factors syntesized by human lung mast cells exert angiogenic effects. J Allergy Clin Immunol 123:1142–1149

Devalaraja RM, Nanney LB, Qian Q et al (2000) Delayed wound healing in CXCR2 knockout mice. J Invest Dermatol 115:234–244

Dias S, Boyd R, Balkwill F (1998) IL-12 regulates VEGF and MMPs in a murine breast cancer model. Int J Cancer 78:361–365

Diaz-Flores L, Gutierrez R, Madrid JF et al (2009) Pericytes. Morphofunction, interaction and pathology in a quiescent and activated mesenchymal cell niche. Histol Histopathol 24:909–969

Dietrich J, Wang D, Batchelor TT (2009) Cediranib: profile of a novel anti-angiogenic agent in patients with glioblastoma. Expert Opin Investig Drugs 18:1549–1557

Dhanabal M, Ramchandran R, Waterman MJ et al (1999a) Endostatin induces endothelial cell apoptosis. J Biol Chem 274:11721–11726

Dhanabal M, Ramchandarn R, Volk R et al (1999b) Endostatin: yeast production, mutants, and antitumor effect in renal cell carcinoma. Cancer Res 59:189–197

Dixelius J, Cross M, Matsumoto T et al (2002) Endostatin regulates endothelial cell adhesion and cytoskeletal organization. Cancer Res 62:1944–1947

Djonov V, Andres AC, Ziemiecki A (2001) Vascular remodelling during the normal and malignant life cycle of the mammary gland. Microsc Res Techn 52:182–189

Dome B, Timar J, Dobos J et al (2006) Identification and clinical significance of circulating endothelial progenitor cells in human non-small cell lung cancer. Cancer Res 66:7341–7347

Dong J, Grunstein J, Tejada M et al (2004) VEGF-null cells require PDGFR alpha signaling-mediated stromal fibroblast recruitment for tumorigenesis. EMBO J 23:2800–2810

Donoghue JF, Lederman FL, Susil BJ et al (2007) Lymphangiogenesis of normal endometrium and endometrial adenocarcinoma. Human Reprod 22:1705–1713

Dregdge K, MMarriott JB, Macdonald CD et al (2002) Novel thalidomide analogues display anti-angiogenic activity independently of immunomodulatory effects. Br J Cancer 87:1162–1172

Dubravec DB, Spriggs DR, Mannick JA et al (1990) Circulating human peripheral blood granulocytes synthesize and secrete tumor necrosis factor alpha. Proc Natl Acad Sci U S A 87:6758–6761

Dvorak AM, Mihm MC Jr, Osage JE et al (1980) Melanoma. An ultrastructural study of the host inflammatory and vascular responses. J Invest Dermatol 75:388–393

Dvorak HF, Nagy JA, Dvorak JT et al (1988) Identification and characterization of the blood vessel of solid tumors that are leaky to circulating macromolecules. Am J Pathol 133:95–109

Dvorak HF, Gresser I (1989) Microvascular injury in pathogenesis of interferon-induced necrosis of subcutaneous tumors in mice. J Natl Cancer Inst 81:497–502

Eberhard A, Kahlert S, Goede V et al (2000) Heterogeneity of angiogenesis and blood vessel maturation in human tumors: implications for antiangiogenic tumor therapies. Cancer Res 60:1388–1393

Eberhard A, Kahlert S, Goede V et al (2002) Heterogeneity of angiogenesis and blood vessel maturation in human tumors: implications for antiangiogenic tumor therapies. Cancer Res 60:1388–1393

Ebos JM, Lee CR, Cruz-Munoz W et al (2009) Accelerated metastasis after short-term treatment with a potent inhibitor of tumor angiogenesis. Cancer Cell 15:232–239

Eichmann A, Corbel C, Nataf V et al (1997) Ligand-dependent development of the endothelial and hemopoietic lineages from embryonic mesodermal cells expressing vascular endothelial growth factor receptor-2. Proc Nat Acad Sci U S A 94:5141–5146

El Hallani S, Boisselier B, Peglion F et al (2010) A new alrenative mechanism in glioblastoma vascularization: tubular vasculogenic mimicry. Brain 133:973–982

Elayadi AN, Samli KN, Prudkin L et al (2007) A peptide selected by biopanning identifies the integrin alphavbeta6 as a prognostic biomarker for nonsmall cell lung cancer. Cancer Res 67:5889–5895

Ellis LM, Staley CA, Liu W et al (1998) Downregulation of vascular endothelial growth factor in a human colon carcinoma cell line transfected with an antisense expression vector specific for c-src. J Biol Chem 273:1052–1057

Ellis LM, Hicklin DJ (2008) Pathways mediating resistance to vascular endothelial growth factor targeted therapy. Clin Cancer Res 14:6371–6375

Elpek GO, Gelen T, Aksoy NH et al (2001) The prognostic relevance of angiogenesis and mast cells in squamous cell carcinoma of the oesophagus. J Clin Pathol 54:940–944

Engsig MT, Chen QJ, Vu TH et al (2000) Matrix metalloproteinase-9 and vascular endothelial growth factor are essential for osteoclast recruitment into developing long bones. J Cell Biol 151:879–890

Erber R, Thurnher A, Katsen AD et al (2004) Combined inhibition of VEGF and PDGF signaling enforces tumor vessel regression by interfering with pericyte-mediated endothelial cell survival mechanisms. FASEB J 18:338–340

Ellis PM, Kaiser R, Zhao Y et al (2010) Phase I open-label study of continuous treatment with BIBF 1120, a triple angiokinase inhibitor, and pemetrexed in pretreated non-small cell lung cancer patients. Clin Cancer Res 16:2281–2289

Escudier B, Eisen T, Stadler WM et al (2007) Sorafenib in advanced clear-cell renal-cell carcinoma. N Engl J Med 356:125–134

Escudier B, Szczylik C, Hutson TE et al (2009) Randomized phase II trial of first-line treatment with sorafenib versus interferon alpha-2 in patients with metastatic renal cell carcinoma. J Clin Oncol 27:1280–1289

Ezekowitz RA, Mulliken JB, Folkman J (1992) Interferon alfa-2a therapy for lifethreatening hemangiomas of infancy. N Engl J Med 326:1456–1463

Falcon BL, Hashizume H, Koumoutsakos P et al (2009) Contrasting actions of selective inhibitors of angiopoietin-1 and angiopoietin-2 on the normalization of tumor blood vessels. Am J Pathol 175:2159–2170

Feistritzer C, Kaneider NC, Sturn DH et al (2004) Expression and function of the vascular endothelial growth factor receptor FLT-1 in human eosinophils. Am J Respir Cell Mol Biol 30:729–735

Feldman DR, Baum M, Ginsberg MS et al (2009) Phase I trial of bevacizumab plus escalated doses of sunitinib in patients with metastatic renal cell carcinoma. J Clin Oncol 27:1432–1439

Fenzl A, Schopmann SF, Geleff S et al (2006) Vascular endothelial growth factor-C expression and lymphangiogenesis in colorectal cancer. Eur Surg 38:149–154

Fernandez A, Udagawa T, Schwesinger C et al (2001) Angiogenic potential of prostate carcinoma cells overexpressing bcl-2. J Natl Cancer Inst 93:208–213

Fernandez Pujol B, Lucibello FC et al (2000) Endothelial-like cells derived from human CD14 positive monocytes. Differentiation 65:287–300

Fidler IJ (1995) Modulation of the organ microenvironment for treatment of cancer metastasis. J Natl Cancer Inst 87:1588–1592

Fidler IJ, Ellis LM (1994) The implications of angiogenesis for the biology and therapy of cancer metastasis. Cell 79:185–188

Fokt RM, Templeton A, Gillesen S et al (2009) Prostatic metastasis of renal cell carcinoma successfully treated with sunitinib. Urol Int 83:122–124

Folkman J (1971) Tumor angiogenesis: therapeutic implications. N Engl J Med 285:1182–1186

Folkman J, Langer R, Linhardt R et al (1983) Angiogenesis inhibition and tumor regression caused by heparin or a heparin fragment in the presence of cortisone. Science 221:719–725

Folkman J, Jalluri R (2003) Tumor angiogenesis. In: Kufe DW, Pollock RE, Weichselbaum RR et al (ed) Cancer medicine. BC Decker, Hamilton, pp 161–194

Fonsato V, Buttigleir S, Deregibus M et al (2006) Expression of PAX2 in human renal tumor-derived endothelial cells sustains apoptosis resistance and angiogenesis. Am J Pathol 168:706–713

Fox WD, Higgins B, Maiese KM et al (2002) Antibody to vascular endothelial growth factor slows growth of an androgen-independent xenograft model of prostate cancer. Clin Cancer Res 8:3226–3231

Fukumoto S, Morifuji M, Katakura Y (2005) Endostatin inhibits lymph node metastasis by a down-regulation of the vascular endothelial growth factor C expression in tumor cells. Clin Exp Metastasis 22:31–38

Fukumura D, Duda DG, Munn LL et al (2010) Tumor microvasculature and microenvironment: novel insights through intravital imaging in pre-clinical models. Microcirculation 17:206–225

Fukunaga M (2005) Expression of D2-40 in lymphatic endothelium of normal tissues and in vascular tumours. Histopathology 46:396–402

Fukushima N, Satoh T, Sano M et al (2001) Angiogenesis and mast cells in non-Hodgkin's lymphoma: a strong correlation in angioimmunoblastic T-cell lymphoma. Leuk Lymphoma. 42:709–720

Furstenberger G, von Moos R, Lucas R et al (2006) Circulating endothelial cells and angiogenic serum factors during neoadjuvant chemotherapy of primary breast cancer. Br J Cancer 94: 524–531

Gabizon A, Catane R, Uziely B et al (1994) Prolonged circulation time and enhanced accumulation in malignant exudates of doxorubicin encapsulated in polyethylene-glycol coated liposomes. Cancer Res 54:987–992

Gabizon A, Isacson R, Rosengarten O et al (2008) An open-label study to evaluate dose and cycle dependence of the pharmacokinetics of pegylated liposomal doxorubicin. Cancer Chemother Pharmacol 61:695–702

Gao D, Nolan DJ, Mellick AS et al (2008) Endothelial progenitor cells control the angiogenic switch in mouse lung metastasis. Science 319:195–198

Gao P, Zhou GY, Zheng QH et al (2009) Lymphangiogenesis in gastric carcinoma correlates with prognosis. J Pathol 218: 192–200.

Garcia-Barros M, Paris F, Cordon-Cardo C et al (2003) Tumor response to radiotherapy regulated by endothelial cell apoptosis. Science 300:1155–1159

Garde SV, Forte AJ, Ge M et al (2007) Binding and internalization of NGR-peptide-targeted liposomal doxorubicin (TVT-DOX) in CD13-expressing cells and its antitumor effects. Anticancer Drugs 18:1189–1200

Garofalo A, Naumova E, Manenti L et al (2003) The combination of the tyrosine kinase receptor inhibitor SU6668 with paclitaxel affects ascites formation and tumor spread in ovarian carcinoma xenografts growing orthotopically. Clin Cancer Res 9:3476–3485

Gauler T, Fischer B, Soria JC et al (2006) Phase II open-label study to investigate the efficacy and safety of PTK787/ZK222584 orally administered once daily at 1,250 mg as a second-line monotherapy in patients with stage IIIB or stage IV non-small cell lung cancer. J Clin Oncol 24:7195

Gehlin UM, Ergun S, Schumacher U et al (2000) In vitro differentiation of endothelial cells from AC 133 positive progenitor cells. Blood 95:3106–3112

Geng L, Donnelly E, McMahon G et al (2001) Inhibition of vascular endothelial growth factor receptor signaling leads to reversal of tumor resistance to radiotherapy. Cancer Res 61:2413–2419

Gerhardt H, Semb H (2008) Pericytes: gatekeepers in tumor cell metastasis? J Mol Med 86:135–144

Gerszten RE, Garcia-Zapeda EA, Lim YC et al (1999) MCP-1 and IL-8 trigger firm adhesion of monocytes to vascular endothelium under flow conditions. Nature 398:718–723

Giatromanolaki A, Sividis E, Minopoulos G et al (2002) Differential assessment of vascular survival ability and tumor angiogenic activity in colorectal cancer. Clin Cancer Res 8:1185–1191

Ginns LC, Roberts DH, Mark EJ et al (2003) Pulmonary capillary hemangiomatosis with atypical endotheliomatosis: successul antiangiogenic therapy with doxycycline. Chest 124:2017–2022

Glade Bender J, Cooney EM, Kandel JJ et al (2004) Vascular remodeling and clinical resistance to antiangiogenic cancer therapy. Drug Resist Update 7:289–300

Glowacki J, Mulliken JB (1982) Mast cells in hemangiomas and vascular malformations. Pediatrics 70:48–51

Go RS, Jobe DA, Asp KE et al (2008) Circulating endothelial cells in patients with chronic lymphocytic leukemia. Ann Hematol 87:369–373

Gombos Z, Xu X, Chu CS et al (2005) Peritumoral lymphatic vessel density and vascular endothelial growth factor C expression in early stage squamous cell carcinoma of the uterine cervix. Clin Cancer Res 11:8367–8371

Gomez-Rivera F, Santillan-Gomez AA, Younes MN et al (2007) The tyrosine kinase inhibitor, AZD2171, inhibits vascular endothelial growth factor receptor signaling and growth of anaplastic thyroid cancer in an orthotopic nude mouse model. Clin Cancer Res 13:4519–4527

Good DJ, Polverini PJ, Rastinejad F et al (1990) A tumor-suppressor dependent inhibitor of angiogenesis is immunologically and functionally indistinguishable from a fragment of thrombospondin. Proc Natl Acad Sci U S A 87:6624–6628

Goodlad RA, Ryan AJ, Wedge SR et al (2006) Inhibiting vascular endothelial growth factor receptor-2 signaling reduces tumor burden in the ApcMin/+ mouse model of early intestinal cancer. Carcinogenesis 27:2133–2139

Gordon AN, Granai CO, Rose PG et al (2000) Phase II study of liposomal doxorubicin in platinum- and paclitaxel-refractory epithelial ovarian cancer. J Clin Oncol 18:3093–3100

Goss G, Arnold A, Shepherd F et al (2010) Randomized, double-blind trial of carboplatin and paclitaxel with either oral cediranib or placebo in advanced non-small cell lung cancer: NCIC clinical trials group BR24 study. J Clin Oncol 28:49–55

Gottfried E, Kreutz M, Haffner S et al (2007) Differentiation of human tumour-associated dendritic cells into endothelial-like cells: an alternative pathway of tumour angiogenesis. Scand J Immunol 65:329–335

Graeber TG, Osmanian C, Jacks T et al (1996) Hypoxia-mediated selection of cells with diminished apoptotic potential in solid tumours. Nature 379:88–91

Graham R, Graham J (1996) Mast cells and cancer of the cervix. Surg Gynecol Obstet 123:3–9

Grant MB, May WS, Caballero S et al (2002) Adult hematopoietic stem cells provide functional hemangioblast activity during retinal neovascularization. Nat Med 8:607–612

Grenier A, Chollet-Martin S, Crestani B et al (2002) Presence of a mobilizable intracellular pool of hepatocyte growth factor in human polymorphonuclear neutrophils. Blood 99:2997–3004

Griscelli F, Li H, Cheong C et al (2000) Combined effects of radiotherapy and angiostatin gene therapy in glioma tumor model. Proc Natl Acad Sci U S A 97:6698–6703

Gruber BL, Marchese MJ, Kew R (1995) Angiogenic factors stimulate mast cell migration. Blood 86:2488–2493

Grunewald M, Avraham I, Dor Y et al (2006) VEGF-induced adult neovascularization: recruitment, retention, and role of accessory cells. Cell 124:175–189

Grützkau A, Krüger-Krasagakes S, Baumeister H et al (1998) Synthesis, storage, and release of vascular endothelial growth factor/vascular permeability factor (VEGF/VPF) by human mast cells: implications for the biological significance of VEGF206. Mol Biol Cell 9:875–884

Gulati R, Jevremovic D, Peterson TE et al (2003) Diverse origin and function of cells with endothelial phenotype obtained from adult human blood. Circ Res 93:1023–1025

Gunsilius E, Duba HC, Petzer A et al (2000) Evidence from a leukaemia model for maintenance of avscular endothelium by bone-marrow-derived endothelial cells. Lancet 355:1688–1691

Guo B, Zhang Y, Luo G et al (2009) Lentivirus-mediated small interfering RNA targeting VEGF-C inhibited tumor lymphangiogenesis and growth in breast carcinoma. Anat Rec 292:633–639

Guo P, Hu B, Gu W et al (1985) Platelet-derived growth factor-B enhances glioma angiogenesis by stimulating vascular endothelial growth factor expression in tumor endothelia and by promoting pericyte recruitment. Am J Pathol 162:1083–1093

Gutschalk CM, Herold-Mende CC, Fusenig NE et al (2006) Granulocyte colony-stimulating factor and granulocyte-macrophage colony stimulating factor promote malignant growth of cells from head and neck squamous cells carcinomas in vivo. Cancer Res 66:8026–8036

Hagemann T, Biswas SK, Lawrence T et al (2009) Regulation of macrophage function in tumors: the multifaceted role of NF-kappa B. Blood 113:3139–3146

Hainsworth JD, Spigel DR, Sosman JA et al (2007) Treatment of advanced renal cell carcinoma with the combination bevacizumab/erlotinib/imatinib: a phase I/II trial. Clin Genitourin Cancer 5:427–432

Hamzah J, Jugold M, Kiessling F et al (2008) Vascular normalization in Rgs5-deficient tumours promotes immune destruction. Nature 453:410–414

Hanada T, Nakagawa M, Emoto A et al (2000) Prognostic value of tumor-associated macrophage count in human bladder cancer. Int J Urol 7:263–269

Hanahan D (1985) Heritable formation of pancreatic beta-cell tumors in transgenic mice expressing recombinant unsylin/simian virus 40 oncogene. Nature 315:115–122

Hanahan D, Bergers G, Bergsland E (2000) Less is more, regularly: metronomic dosing of cytotoxic drugs can target tumor angiogenesis in mice. J Clin Invest 105:1045–1047

Haniffa MA, Collin MP, Buckley CD et al (2009) Mesenchymal stem cells: the fibroblasts' new clothes? Haematoloica 94:258–263

Hartveit F (1981) Mast cells and metachromasia in human breast cancer: their occurrence, significance and consequence: a preliminary report. J Pathol 134:7–11

Hashizume H, Baluk P, Morikawa S et al (2000) Openings between defective endothelial cells explain tumor vessel leakiness. Am J Pathol 156:1363–1380

Hattori K, Dias S, Heissig B et al (2000) Vascular endothelial growth factor and angiopoietin-1 stimulate postnatal hematopoiesis by recruitment of vasculogenic and hematopoietic stem cells. J Exp Med 193:1005–1014

Hattori K, Dias S, Heissig B et al (2001) Vascular endothelial growth factor and angiopoietin-1 stimulate postnatal hematopoiesis by recruitment of vasculogenic and hematopoietic stem cells. J Exp Med 193:1005–1014

He XW, Liu T, Xiao Y et al (2009) Vascular endothelial growth factor-C siRNA delivered via calcium carbonate nanoparticle effectively inhibits lymphangiogenesis and growth of colorectal cancer in vivo. Cancer Biother Radiopharm 24:249–259

He Y, Kozaki K, Karpanen T et al (2002) Supression of tumor lymphangiogenesis and lymph node metastasis by blocking vascular endothelial growth factor receptor 3 signalling. J Natl Cancer Inst 94:819–825

He Y, Rajantie I, Ilmonen M et al (2004) Preexisting lymphatic endothelium but not endothelial progenitor cells are essential for tumor lymphangiogenesis and lymphatic metastasis. Cancer Res 64:3737–3740

He Y, Rajantie, I, Pajusola K et al (2005) Vascular endothelial cell growth factor receptor 3-mediated activation of lymphatic endothelium is crucial for tumor cell entry and spread via lymphatic vessels. Cancer Res 65:4739–4746

Heeschen C, Aicher A, Lehmann R et al (2003) Erythropoietin is a potent physiologic stimulus for endothelial progenitor cell mobilization. Blood 102:1340–1346

Heilberg C, Ostman A, Heldin CH (2010) PDGF-C and vessel maturation. Rec Res Cancer Res 180:103–114

Helfrich I, Scheffrahn I, Bartling S et al (2010) Resistance to antiangiogenic therapy is directed by vascular phenotype, vessel stabilization, and maturation in malignant melanoma. J Exp Med 207:491–503

Hendrix MJ, Seftor EA, Mettzer PS et al (2001) Expression and functional significance of VE-cadherin in aggressive human melanoma cells: role in vasculogenic mimicry. Proc Natl Acad Sci U S A 98:8018–8023

Herbst R, Sun Y, Korfee S et al (2009) Vandetanib plus docetaxel versus docetaxel as second-line treatment for patients with advanced non-small cell lung cancer (NSCLC): a randomized, double-blind phase III trial (ZODIAC). J Clin Oncol 27 (Abstract CRA8003)

Hess AR, Seftor EA, Gardner LM et al (2001) Molecular regulation of tumor cell vasculogenic mimicry by tyrosine phosphorylation: role of epithelial cell kinase (Eck/EphA2). Cancer Res 61:3250–3255

Hess C, Vuong V, Hegyi I et al (2001) Effect of VEGF receptor inhibitor PTK787/ZK222584 (correction of ZK222548) combined with ionizing radiation on endothelial cells and tumour growth. Br J Cancer 85:2010–2016

Heymach JV, Johnson BE, Prager D et al (2007) Randomized, placebo-controlled phase II study of vandetanib plus docetaxel in previously treated non small-cell lung cancer. J Clin Oncol 25:4270–4277

Heymach JV, Paz-Ares L, De Braud F et al (2008) Randomized phase II study of vandetanib alone or with paclitaxel and carboplatin as first-line treatment for advanced non-small-cell lung cancer. J Clin Oncol 26:5407–5415

Hida K, Klagsbrun M (2005) A new perspective of tumor endothelial cells: unexpected chromosome and chromosome abnormalities. Cancer Res 65:2507–2510

Hida K, Hida Y, Amin DN et al (2004) Tumor-associated endothelial cells with cytogenetic abnormalities. Cancer Res 64:8249–8255

Hilberg F, Roth GJ, Krssak M et al (2008) BIBF 1120: triple angiokinase inhibitor with sustained receptor blockade and good antitumor efficacy. Cancer Res 68:4774–4782

Hirakawa S, Kodama S, Kunstfeld R et al (2005) VEGF-A induces tumor and sentinel lymph node lymphangiogenesis and promotes lymphatic metastasis. J Exp Med 201:1089–1099

Hirakawa S, Brown LF, Kodama S et al (2007) VEGF-C-induced lymphangiogenesis in sentinel lymph nodes promotes metastasis to distant sites. Blood 109:1010–1017

Hlatky L, Hahnfeldt P, Folkman J (2002) Clinical application of antiangiogenic therapy: microvessel density, what it does and doesn't tell us. J Natl Cancer Inst 94:883–893

Hlushchuk R, Riesterer O, Baum O et al (2008) Tumor recovery by angiogenic switch from sprouting to intussusceptive angiogenesis after treatment with PTK787/ZK222584 or ionizing radiation. Am J Pathol 173:1173–1185

Hobbs SK, Monsky WL, Yuan F et al (1998). Regulation of transport pathways in tumor vessels: role of tumor type and microenvironment. Proc Natl Acad Sci U S A 95:4607–4612

Holash J, Maisonpierre PC, Compton D et al (1999) Vessel cooption, regression and growth in tumors mediated by angiopoietins and VEGF. Science 284:1994–1998

Holash J, Davis S, Papadopoulos N et al (2002) VEGF-Trap: a VEGF blocker with potent antitumor effects. Proc Natl Acad Sci U S A 99:11393–11398

Holden SN, Eckhardt SG, Basser R et al (2005) Clinical evaluation of ZD6474, an orally active inhibitor of VEGF and EGF receptor signaling, in patients with solid, malignant tumors. Ann Oncol 16:1391–1397

Horiuchi T, Weller PF (1997) Expression of vascular endothelial growth factor by human eosinophils: upregulation by granulocyte macrophage colony-stimulating factor and interleukin-5. Am J Respir Cell Mol Biol 17:70–77

Horsman MR, Siemann DW (2006) Pathophysiologic effects of vascular-targeting agents and the implications for combination with conventional therapies. Cancer Res 66:11520–11539

Hoshino M, Takahashi M, Aoike N (2001) Expression of vascular endothelial growth factor, basic fibroblast growth factor, and angiogenin immunoreactivity in asthmatic airways and its relationship to angiogenesis. Allergy Clin Immunol 107:295–301

Huang PY, Pan Q, Li AI et al (2006) Gr-1[+] CD115[+] immature myeloid suppressor cells mediate the development of tumor-induced T regulatory cells and T-cell anergy in tumor-bearing host. Cancer Res 66:1123–1131

Huang X, Molema G, King S et al (1997) Tumor infarction in mice by antibody-directed targeting of tissue factor to tumor vasculature. Science 275:547–550

Hu-Lowe DD, Zou HY, Grazzini ML et al (2008) Non-clinical antiangiogenesis and antitumor activities of axitinib (AG-013736), an oral, potent, and selective inhibitor of vascular endothelial growth factor receptor tyrosine kinase 1, 2, 3. Clin Cancer Res 14:7272–7283

Huber PE, Bischof M, Jenne J et al (2005) Trimodal cancer treatment: beneficial effects of combined antiangiogenesis, radiation, and chemotherapy. Cancer Res 65:3643–3655

Hunt TK, Aslam T, Hussain Z et al (2008) Lactate, with oxygen, incites angiogenesis. Adv Exp Med Biol 614:73–80

Hur J, Yoon CH, Kim HS et al (2004) Characterization of two types of endothelial progenitor cells and their different contributions to neovasculogenesis. Arterioscler Thromb Vasc Biol 24:288–293

Hurwitz H, Fehrenbacker K, Novotny W et al (2004) Bevacizumab plus irinotecan, fluorouracil, and leucovorin for metastatic colorectal cancer. N Engl J Med 350:2335–2342

Iba O, Matsubara H, Nozawa Y et al (2002) Angiogenesis by implantation of peripheral blood mononuclear cells and platelets into ischemic limbs. Circulation 106:2019–2025

Igreja C, Courinha M, Cachaco AS et al (2007) Characterization and clinical relevance of circulating and biopsy-derived endothelial progenitor cells in lymphoma patients. Haematologica 92:469–477

Imada A, Shijubo N, Koijma H et al (2000) Mast cells correlate with angiogenesis and poor outcome in stage I lung adenocarcinoma. Eur Resp J 15:1087–1093

Imai K, Takaoka A (2006) Comparing antibody and small-molecule therapies for cancer. Nat Rev Cancer 6:714–727

Inai T, Mancuso M, Hashizume H et al (2004) Inhibition of vascular endothelial growth factor (VEGF) signaling in cancer causes loss of endothelial fenestrations, regression of tumor vessels and appearance of basement membrane ghosts. Am J Pathol 165:35–52

Ingber DE, Madri JA, Folkman J (1986) A possible mechanism for inhibition of angiogenesis by angiostatic steroids: induction of capillary basement membrane dissolution. Endocrinology 119:1768–1775

Ingber D, Fujita T, Kishimoto S, Sudo K, Kanamaro T, Brem H et al (1990) Synthetic analogues of fumagillin that inhibit angiogenesis and suppress tumour growth. Nature 468:555–557

Inoue K, Chikazawa M, Fukata S et al (2003) Docetaxel enhances the therapeutic effect of the angiogenesis inhibitor TNP-470 (AGM-1470) in metastatic human transitional cell carcinoma. Clin Cancer Res 9:886–899

Inoue A, Moriya H, Katada N et al (2008) Intratumoral lymphangiogenesis of esophageal squamous cell carcinoma and relationship with regulatory factors and prognosis. Pathol Int 58:611–619

Iruela-Arispe ML, Bornstein P, Sage H (1991) Thrombospondin exerts an antiangiogenic effect on cord formation by endothelial cells in vitro. Proc Natl Acad Sci U S A 88:5026–5030

Isaka N, Padera TP, Hagendoorn J et al (2004) Peritumoral lymphatics induced by vascular endothelial growth factor-C exhibit abmormal function. Cancer Res 64:4400–4404

Isner JM, Asahara T (1999) Angiogenesis and vasculogenesis as therapeutic strategies for postnatal neovascularization. J Clin Invest 103:1231–1236

Italiano JE Jr, Richardson JL, Paltel-Hett S et al (2008) Angiogenesis is regulated by a novel mechanism: pro- and antiangiogenic proteins are organized into separate platelet alpha granules and differentially released. Blood 111:1227–1233

Ito H, Rovira II, Bloom ML et al (1999) Endothelial progenitor cells as putative targets for angiostatin. Cancer Res 59:5875–5877

Iwaguro H, Yamaguchi J, Kalka C et al (2002) Endothelial progenitor cell vascular endothelial growth factor gene transfer for vascular regeneration. Circulation 105:732–738

Jackson KA, Majka SM, Wang H et al (2001) Regeneration of ischemic cardiac muscle and vascular endothelium by adult stem cells. J Clin Invest 107:1395–1402

Jain RK (1998) The next frontier of molecular medicine: delivery of therapeutics. Nat Med 4:655–657

Jain RK (2001) Normalizing tumor vasculature with anti-angiogenic therapy: a new paradigm for combination therapy. Nat Med 7:987–989

Jain RK (2005) Normalization of tumor vasculature: an emerging concept in antiangiogenic therapy. Science 307:58–62

Jain RK, Fenton BT (2002) Intratumoral lymphatic vessels: a case of mistaken identity or malfunction? J Natl Cancer Inst 94:417–421

Jain RK, Duda DG, Clark JW et al (2006) Lessons from phase III clinical trials on anti-VEGF therapy for cancer. Nat Clin Pract Oncol 3:24–40

Jenkins DC, Charles IG, Thomsen LL et al (1995) Roles of nitric oxide in tumor growth. Proc Natl Acad Sci U S A 92:4392–4396

Ji RC (2006) Lymphatic endothelial cells, tumor lymphangiogenesis and metastasis: new insights into intratumoral and peritumoral lymphatics. Cancer Metastasis Rev 25:677–694

Ji RC, Eshita Y, Kato S (2007) Investigation of intratumoural and peritumoral lymphatics expressed by podoplanin and LYVE-1 in the hybridoma-induced tumours. Int J Exp Pathol 88:257–270

Jin DK, Shido K, Kopp HG et al (2006) Cytokine-mediated deployment of SDF-1 induces revascularization through recruitment of CXCR4$^+$ hemangiocytes. Nat Med 12:557–567

Jones N, Iljin K, Dumont DJ, Alitalo K (2001) Tie receptors: new modulators of angiogenesis and lymphangiogenic responses. Nat Rev Mol Cell Biol 2:257–267

Joensuu H, De Braud F, Coco P et al (2008) Phase II, open-label study of PTK787/ZK222584 for the treatment of metastatic gastrointestinal stromal tumors resistant to imatinib mesylate. Ann Oncol 19:173–177

Johnson DH, Fehrenbacher L, Novotny WF et al (2004) Randomized phase II trial comparing bevacizumab plus carboplatin and paclitaxel with carboplatin and paclitaxel alone in previously untreated locally advanced or metastatic non-small-cell lung cancer. J Clin Oncol 22:2184–2191

Kaban LB, Mulliken JB, Ezekowiyz RA et al (1999) Antiangiogenic therapy of a recurrent giant cell tumor of the mandible with interferon alpha-2a. Pediatrics 103:1145–1149

Kaban LB, Troulis MJ, Ebb D et al (2002) Antiangiogenic therapy with interferon alpha for giant cell lesions of the jaws. J Oral Maxillofac Surg 60:1103–1111

Kabbinavar F, Hurwitz HI, Fehrenbacher L et al (2003) Phase II, randomized trial comparing bevacizumab plus fluorouracil (FU) leucovorin (LV) with FU/LV alone in patients with metastatic colorectal cancer. J Clin Oncol 21:60–65

Kaifi JT, Yekebas EF, Schurr P et al (2005) Tumor cell homing to lymph nodes and bone marrow and CXCR4 expression in esophageal cancer. J Natl Cancer Inst 97:1840–1847

Kalka C, Masuda H, Takahashi T et al (2000) Transplantation of ex vivo expanded endothelial progenitor cells for therapeutic neovascularization. Proc Nat Acad Sci U S A 97:3422–3427

Kalluri R (2003) Basement membranes: structure, assembly and role in tumour angiogenesis. Nat Rev Cancer 3:422–433

Kamen BA, Glod J, Cole PD (2006) Metronomic therapy from a pharmacologist's view. J Pediatr Hematol Oncol 28:325–327

Kanbe N, Kurosawa M, Nagata H et al (2000) Production of fibrogenic cytokines by cord blood-derived cultured human mast cells. J Allergy Clin Immunol 106:S85–S90

Kaneko I, Tanaka S, Oka S et al (2006) Lymphatic vessel density at the site of deepest penetration as a predictor of lymph node metastasis in submucosal colorectal cancer. Dis Colon Rectum 50:13–21

Karashima T, Inoue K, Fukata S et al (2007) Blockade of the vascular endothelial growth factor-receptor 2 pathway inhibits the growth of human renal cell carcinoma, RBM1-IT4, in the kidney but not in the bone of nude mice. Int J Oncol 30:937–945

Karpanen T, Egeblad M, Karkkainen et al (2001) Vascular endothelial growth factor C promotes tumor lymphangiogenesis and intralymphatic tumor growth. Cancer Res 61:1786–1790

Kashiwagi S, Tsukada K, Xu L et al (2008) Perivascular nitric oxide gradients normalize tumor vasculature. Nature Med 14:255–257

Kasibhatla M, Steinberg P, Meyer J et al (2007) Radiation therapy and sorafenib: clinical data and rationale for the combination in metastatic renal cell carcinoma. Clin Genitourin Cancer 5:291–294

Kato Y, Kaneko M, Sata M et al (2005) Enhanced expression of Aggrus (T1α/podoplanin), a platelet aggregation-inducing factor in lung squamous cell carcinoma. Tumor Biol 26:195–200

Kato Y, Kaneko MK, Kuno A et al (2006) Inhibition of tumor cell-induced platelet aggregation using a novel anti-podoplanin antibody reacting with its platelet-aggregation-stimulating domain. Biochem Biophys Res Commun 349:1301–1307

Kato Y, Kaneko MK, Kunita A et al (2007) Molecular analysis of the pathophysiological binding of the platelet aggregation-inducing factor podoplanin to the C-type lectin-like receptor CLEC-2. Cancer Sci 3:1–8

Kawai H, Minamiya Y, Ito M et al (2008) VEGF121 promotes lymphangiogenesis in the sentinel lymph nodes of non-small cell lung carcinoma patients. Lung Cancer 59:41–47

Khanna C, Jaboin JJ, Drakos E et al (2002) Biologically relevant orthotopic neuroblastoma xenograft models: primary adrenal tumor growth and spontaneous distant metastasis. In Vivo 16:77–85

Kerbel RS, Yu J, Tran J et al (2001) Possible mechanisms of acquired resistance to anti-angiogenic drugs: implications for the use of combination therapy approaches. Cancer Metastasis Rev 20:79–86

Kim YM, Hwang S, Pyun BJ et al (2002) Endostatin blocks vascular endothelial growth factor signaling via direct interaction with KDR/Flk-1. J Biol Chem 277:27872–27879

Kimura N, Kimura I (2005) Podoplanin as a marker for mesothelioma. Pathol Int 55:83–86

Kisker O, Becker CM, Prox D et al (2001) Continuous administration of endostatin by intraperitoneally implanted osmotic pump improves the efficacy and potency of therapy in a mouse xenograft tumor model. Cancer Res 61:7669–7774

Kita H, Ohnishi T, Okubo Y et al (1991) Granulocyte/macrophage colony-stimulating factor and interleukin 3 release from human peripheral blood eosinophils and neutrophils. J Exp Med 174:745–748

Klement G, Baruchel S, Rak J et al (2000) Continuous low-dose therapy with vinblastine and VEGF receptor-2 antibody induces sustained tumor regression without overt toxicity. J Clin Invest 105:R15–R24

Klement GL, Yip TT, Cassiola F et al (2009) Platelets actively sequester angiogenesis regulators. Blood 113:2835–2842

Klimp AH, Hollema H, Kempinga C et al (2001) Expression of cyclooxygenase-2 and inducible nitric oxide synthase in human ovarian tumors and tumor-associated macrophages. Cancer Res 61:7305–7309

Kobayashi S, Kishimoto T, Kamata S et al (2007) Rapamycin, a specific inhibitor of the mammalian target rapamycin, suppresses lymphangiogenesis and lymphatic metastasis. Cancer Sci 98:726–733

Koide N, Nishio A, Sato T et al (2004) Significance of macrophage chemoattractant protein-1 expression and macrophage infiltration in squamous cell carcinoma of the esophagus. Am J Gastroenterol 99:1667–1674

Kolonin MG, Pasqualini RW, Arap W (2001) Molecular addresses in blood vessel as targets for therapy. Curr Opin Chem Biol 5:308–313

Koukourakis MI, Giatromanolaki A, Sivridis E et al (2005) LYVE-1 immunohistochemical assessment of lymphangiogenesis in endometrial and lung cancer. J Clin Pathol 58:202–206

Koutras AK, Krikelis D, Alexandrou N et al (2007) Brain metastasis in renal cell cancer responding to sunitinib. Anticancer Res 27:4255–4257

Krishnan J, Kirkin V, Steffen A et al (2003) Differential in vivo and in vitro expression of vascular endothelial growth factor (VEGF)-C and VEGF-D in tumors and its relationship to lymphatic metastasis in immunocompromized rats. Cancer Res 63:713–722

Kujawski M, Kortylewski M, Lee H et al (2008) Stat3 mediates myeloid cell-dependent tumor angiogenesis in mice. J Clin Invest 18:3367–3677

Kunkel P, Ulbricht U, Bohlen P et al (2001) Inhibition of glioma angiogenesis and growth in vivo by systemic treatment with a monoclonal antibody against vascular endothelial growth factor receptor-2. Cancer Res 61:6624–6628

Lachter J, Stein M, Lichting C et al (1995) Mast cells in colorectal neoplasias and premalignant disorders. Dis Colon Rectum 38:290–293

Langer R, Cohn M, Vacanti J et al (1980) Control of tumor growth in animals by infusion of an angiogenesis inhibitor. Proc Natl Acad Sci U S A 77:4331–4335

Lawler J (2002) Thrombospondin-1 as an endogenous inhibitor of angiogenesis and tumor growth. J Cell Mol Med 6:1–12

Leali D, Dell'Era P, Stabile H et al (2003) Osteopontin (Eta-1) and fibroblast growth factor-2 cross-talk in angiogenesis. J Immunol 171:1085–1093

Lee SH, Schloss DJ, Jarvis L et al (2001) Inhibition of angiogenesis by mouse Sprouty protein. J Biol Chem 276:4128–4133

Lee YC, Kwak YG, Song CH (2002) Contribution of vascular endothelial growth factor to airway hyperresponsiveness and inflammation in a murine model of toluene diisocyanate-induced asthma. J Immunol 168:3595–3600

Lee JT, McCubrey JA (2003) BAY-43–9006 Bayer/Onyx. Curr Opin Investig Drugs 4:757–763

Leenders WP, Kusters B, Verrijp K et al (2004) Antiangiogenic therapy of cerebral melanoma metastases results in sustained tumor progression via vessel co-option. Clin Cancer Res 10:6222–6230

Leek RD, Lewis CE, Whitehouse R et al (1996) Association of macrophage infiltration with angiogenesis and prognosis in invasive breast carcinoma. Cancer Res 56:4625–4629

Leek RD, Landers RJ, Harris AL et al (1999) Necrosis correlates with high vascular density and focal macrophage infiltration in invasive carcinoma of the breast. Br J Cancer 79:991–995

Levy AP, Levy NS, Wegner S et al (1995) Transcriptional regulation of the rat vascular endothelial growth factor gene by hypoxia. J Biol Chem 270:13333–13340

Lewis CE, Leek R, Harris A et al (1995) Cytokine regulation of angiogenesis in breast cancer: the role of tumor-associated macrophages. J Leukoc Biol 57:747–751

Lewis CE, Pollard JW (2006) Distinct role of macrophages in different tumor microenvironments. Cancer Res 66:605–612

Li M, Gu Y, Zhang Z et al (2010) Vasculogenic mimicry: a new prognostic sign of gastric adenocarcinoma. Pathol Oncol Res 16:259–266

Liang P, Hong JW, Ubukata H et al (2006) Increased density and diameter of lymphatic microvessel correlate with lymph node metastases in early stage invasive colorectal carcinoma. Virchows Arch 448:570–575

Liang WC, Wu X, Peale F et al (2006) Cross-species vegf-blocking antibodies completely inhibit the growth of human xenografts and measure the contribution of stromal VEGF. J Biol Chem 281:951–961

Lin EY, Nguyen AV, Russell RG et al (2001) Colony-stimulating factor 1 promotes progression of mammary tumors to malignancy. J Exp Med 193:727–740

Lin EY, Li JF, Gnatovskiy L, Deng Y et al (2006) Macrophages regulate the angiogenic switch in a mouse model of breast cancer. Cancer Res 66:11238–11246

Lin Y, Weisdorf DJ, Solovey A et al (2000) Origins of circulating endothelial cells and endothelial outgrowth from blood. J Clin Invest 105:71–77

Linhardt R, Grant A, Cooney C et al (1982) Differential anticoagulant activity of heparin fragments prepared using microbial heparinase. J Biol Chem 257:7310–7313

Lissbrant IF, Stattin P, Wikstrom P et al (2000) Tumor associated macrophages in human prostate cancer: relation to clinicopathological variables and survival. Int J Oncol 17:445–451

Liu J, Liao S, Huang Y et al (2011) PDGF-C improves drug delivery and efficacy via vascular normalization, but promotes lymphatic metastasis by activating CXCR4 in breast cancer. Clin Cancer Res 17:3638–3648

Llovet JM, Ricci S, Mazzaferro V et al (2008) Sorafenib in advanced hepatocellular carcinoma. N Engl J Med 359:378–390

Loi M, Marchio' S, Becherini P et al (2010) Combined targeting of perivascular and endothelial tumor cells enhances anti-tumor efficacy of liposomal chemotherapy in neuroblastoma. J Control Release 145:66–73

Lu C, Bonome T, Li Y et al (2007) Gene alterations identified by expression profiling in tumor-associated endothelial cells from invasive ovarian carcinoma. Cancer Res 67:1757–1768

Lu C, Shahzad MM, Moreno-Smith M et al (2010) Targeting pericytes with a PDGF-B aptamer in human ovarian carcinoma models. Cancer Biol Ther 9:176–182

Lyden D, Hattori K, Dias S et al (2001) Impaired recruitment of bone-marrow-derived endothelial and hematopoietic precursor cells blocks tumor angiogenesis and growth. Nature Med 7:1194–1201

Machein MR, Knedla A, Knoth R et al (2004) Angiopoietin-1 promotes tumor angiogenesis in a rat glioma model. Am J Pathol 165:1557–1570

Maciag PC, Seavey MM, Pan ZK et al (2008) Cancer immunotherapy targeting the high molecular weight melanoma-associated antigen protein results in a broad antitumor response and reduction of pericytes in the tumor vasculature. Cancer Res 68:8066–8075

Madden SL, Cook BP, Ncht M et al (2004) Vascular gene expression in nonneoplastic and malignant brain. Am J Pathol 165:601–608

Magnussen A, Kasman IM, Norberg S et al (2005) Rapid access of antibodies to $\alpha5\beta1$ integrin overexpressed on the luminal surface of tumor blood vessels. Cancer Res 65:2712–2721

Mancuso P, Burlini A, Pruneri G et al (2001) Resting and activated endothelial cells are increased in the peripheral blood of cancer patients. Blood 97:3658–3661

Mancuso MR, Davis R, Norberg SM et al (2006) Rapid vascular regrowth in tumors after reversal of VEGF inhibition. J Clin Invest 116:2610–2621

Maniotis AJ, Folberg R, Hess A et al (1999) Vascular channel formation by human melanoma cells in vivo and in vitro: vasculogenic mimicry. Am J Pathol 155:739–752

Mantovani A, Bottazzi B, Colotta F et al (1992) The origin and function of tumor-associated macrophages. Immunol Today 13:265–270

Mantovani A, Sozzani S, Locati M et al (2002) Macrophage polarization: tumor-associated macrophages as a paradigm for polarized M2 mononuclear phagocytes. Trends Immunol 23:549–555

Makitie T, Summanen P, Takkanen A et al (2001) Tumor-infiltrating macrophages (CD68(+) cells) and prognosis in malignant uveal melanoma. Invest Ophtalmol Vis Sci 42:1414–1421

Marchiò S, Lahdenranta J, Schlingemann RO et al (2004) Aminopeptidase A is a functional target in angiogenic blood vessels. Cancer Cell 5:151–162

Marinho VFZ, Metze K, Sanches FSF et al (2008) Lymph vascular invasion in invasive mammary carcinomas identified by the endothelial lymphatic marker D2-40 is associated with other indicators of poor prognosis. BMC Cancer 8:64

Martin V, Liu D, Gomez-Manzano C (2009) Encountering and adavancing through antiangiogenic therapy for gliomas. Curr Pharm Des 15:353–364

Masood R, Gordon EM, Whitley MD et al (2001) Retroviral vectors bearing IgG-binding motifs for antibody-mediated targeting of vascular endothelial growth factor receptors. Int J Mol Med 8:335–343

Massa M, Rosti V, Ramajoli I et al (2005) Circulating CD34^{+}, CD133^{+}, and vascular endothelial growth factor receptor-2 positive endothelial progenitor cells in myelofibrosis with myeloid metaplasia. J Clin Oncol 23:5688–5695

Massi D, Puig S, Franchi A et al (2006) Tumour lymphangiogenesis is a possible predictor of sentinel lymph node status in cutaneous melanoma: a case-control study. J Clin Pathol 59:166–173

Masson V, De La Ballina LR, Munaut C et al (2005) Contribution of host MMP-2 and MMP-9 to promote tumor vascularization and invasion of malignant keratinocytes. FASEB J 19:234–236

Matsumoto K, Nakayama Y, Inoue Y et al (2006) Lymphatic microvessel density is an independent prognostic factor in colorectal cancer. Dis Colon Rectum 50:308–314

Mc Donald DM, Baluk P (2002) Significance of blood vessels leakiness in cancer. Cancer Res 62:5381–5385

Mc Donald DM, Choyke PL (2003) Imaging of angiogenesis: from microscope to clinic. Nat Med 9:713–725

Mc Donald DM, Munn L, Jain RK (2000) Vasculogenic mimicry: how convincing, how novel, and how significant. Am J Pathol 156:383–388

McGee MC, Hammer JB, Willimas RF et al (2010) Improved intratumoral oxygenation through vascular normalization increases glioma sensitivity to ionizing radiation. Int J Radiat Oncol Biol Phys 76:1537–1545

Medioni J, Choueri TK, Zinzindohoue F et al (2009) Response of renal cell carcinoma pancreatic metastasis to sunitinib treatment: a retrospective analysis. J Urol 181:2470–2475

Mellin GW, Katzenstein M (1962) The saga of thalidomide. N Engl J Med 267:1184–1193

Mendel DB, Laird AD, Xin X et al (2003) In vivo antitumor activity of SU11248, a novel tyrosine kinase inhibitor targeting vascular endothelial growth factor and platelet-derived growth factor receptors: determination of a pharmacokinetic/pharmacodynamic relationship. Clin Cancer Res 9:327–337

Metcalfe DD, Baram D, Mekori YA (1997) Mast cells. Physiol Rev 77:1033–1079

Mignatti P, Rifkin DB (1993) Biology and biochemistry of proteinases in tumor invasion. Physiol Rev 73:161–195

Milkowski DM, Weiss RA (1999) TNP 470. In: Teicher BA (ed) Antiangiogenic agents in cancer therapy. Humana, Totowa, pp 385–398

Miller KD (2003) E2100: a phase III trial of paclitaxel versus paclitaxel/bevacizumab for metastatic breast cancer. Clin Breast Cancer 3:421–422

Miller KD, Sweeney CJ, Sledge GW Jr (2003) The snark is a boojum: the continuing problem of drug resistance in the antiangiogenic era. Ann Oncol 14:20–28

Miller KD, Sweeney CJ, Sledge GW Jr (2005) Can tumor angiogenesis be inhibited without resistance? EXS 95–112

Miller KD, Miller M, Mehrotra S et al (2006) A physiologic imaging pilot study of breast cancer treated with AZD2171. Clin Cancer Res 12:281–288

Miller KD, Wang M, Gralow J et al (2007) Paclitaxel plus bevacizumab versus paclitaxel alone for metastatic breast cancer. N Engl J Med 357:2666–2676

Millimaggi D, Mari M, D'Ascenzo S et al (2007) Tumor vesicle-associated CD147 modulates the angiogenic capability of endothelial cells. Neoplasia 9:349–357

Mimura K, Kono K, Takahashi A et al (2007) Vascular endothelial growth factor inhibits the function of human mature dendritic cells mediated by VEGF receptor-2. Cancer Immunol Immunother 56:761–770

Miraglia S, Godfrey W, Yin AH et al (1997) A novel five-transmembrane hematopoietic stem cell antigen: isolation, characterization and molecular cloning. Blood 90:5013–5021

Mishima K, Kato Y, Kaneko MK et al (2006a) Increased expression of podoplanin in malignant astrocytic tumors as a novel molecular marker of malignant progression. Acta Neuropathol 111:483–488

Mishima, K, Kato Y, Kaneko MK et al (2006b) Podoplanin expression in primary central nervous system germ cell tumors: a useful histological marker for the diagnosis of germinoma. Acta Neuropathol 111:563–568

Mishira PJ, Mishra PJ, Humeniuk R et al (2008) Carcinoma-associated fibroblast-like differentiation of human mesenchymal stem cells. Cancer Res 68:4331–4339

Miyahara, M, Tanuma J, Sugihara K et al (2007) Tumor lymphangiogenesis correlates with lymph node metastasis and clinicopathologic parameters in oral squamous cell carcinoma. Cancer 110:1287–1294

Mohammed RAA, Ellis IO, Elsheikh S et al (2009) Lymphatic and angiogenic characteristics in breast cancer: morphometric analysis and prognostic implications. Breast Cancer Res Treat 113:261–274

Mohle R, Bautz F, Rafii S et al (1998) The chemokine receptor CXCR-4 is expressed on CD34[+] hematopoietic progenitors and leukemic cells and mediates transendothelial migration induced by stromal cell-derived factor-1. Blood 91:4523–4530

Moldovan NI, Goldschmidt-Clermont PJ, Parker-Thomburg J et al (2000) Contribution of monocytes/macrophages to compensatory neovascularization: the drilling of metalloelastase-positive tunnels in ischemic myocardium. Circ Res 87:378–384

Molica S, Vacca A, Crivellato E et al (2003) Tryptase-positive mast cells predict clinical outcome of patients with B-cell chronic lymphocytioc leukemia. Eur J Haematol 71:137–139

Moller A, Lippert U, Lessmann D et al (1993) Human mast cells produce IL-8. J Immunol 151:3261–3266

Monestiroli S, Mancuso P, Burlini A et al (2001) Kinetics and viability of circulating endothelial cells as surrogate angiogenesis marker in an animal model of human lymphoma. Cancer Res 61:4341–4314

Morikawa S, Baluk P, Kaido H et al (2002) Abnormalities in pericytes on blood vessels and endothelial sprouts in tumors. Am J Pathol 160:985–1000

Moses MA, Sudhalter J, Langer R (1990) Identification of an inhibitor of neovascularization from cartilage. Science 248:1408–1410

Moss TJ, Reynolds CP, Sather HN (1991) Prognostic value of immunocytologic detection of bone marrow metastases in neuroblastoma. N Engl J Med 324:219–226

Motzer RJ, Rini BI, Bukowski RM et al (2006) Sunitinib in patients with metastatic renal cell carcinoma. JAMA 295:2516–2524

Motzer RJ, Hutson TE, Tomczak P et al (2007) Sunitinib versus interferon alfa in metastatic renal-cell carcinoma. N Engl J Med 356:115–124

Motzer RJ, Hutson TE, Tomczak P et al (2009) Overall survival and updated results for sunitinib compared with interferon alfa in patients with metastatic renal cell carcinoma. J Clin Oncol 27:3584–3590

Mueller MM (2008) Inflammation anf angiogenesis: innate immune cells as modulators of tumor vascularization. In: Marmé D, Fusenig N (ed) Tumor angiogenesis. Basic mechanisms and cancer therapy. Springer. Berlin, pp 351–362

Muhs BE, Plitas G, Delgado Y et al (2003) Temporal expression and activation of matrix metalloproteinases-2, -9, and membrane type 1-matrix metalloproteinase following acute hindlimb ischemia. J Surg Res 111:8–15

Murdoch C, Gannoudis A, Lewis CE (2004) Mechanisms regulating the recruitment of macrophages into hypoxic areas of tumors and other ischemic tissues. Blood 104:2224–2234

Murdoch C, Tazzyman S, Webster S et al (2007) Expression of Tie-2 by human monocytes and their responses to angiopoietin-2. J Immunol 178:7405–7411

Murphy EA, Shields DJ, Stoletov K et al (2010) Disruption of angiogenesis and tumor growth with an orally active drug that stabilzes the inactive state of PDGFRβ/B-RAF. Proc Natl Acad Sci U S A 107:4299–4304

Murray LJ, Abrams TJ, Long, KR et al (2003) SU11248 inhibits tumor growth and CSF-1R-dependent osteolysis in an experimental breast cancer bone metastasis model. Clin Exp Metastasis 20:755–766

Nagy JA, Vasile E, Feng D et al (2002) Vascular permeability factor/vascular endothelial growth factor induces lymphangiogenesis as well as angiogenesis. J Exp Med 196:1497–1506

Nagy JA, Chang SH, Shih SC et al (2010) Heterogeneity of the tumor vasculature. Semin Thromb Hemost 36:321–331

Naik RP, Jin D, Chuang E et al (2008) Circulating endothelial progenitor cells correlate to stage in patients with invasive breast cancer. Breast Cancer Res Treat 107:133–138

Nakao S, Kuwano T, Tsutsumi-Miyahara C et al (2005) Infiltration of COX-2-expressing macrophages is a prerequisite for IL-1 beta-induced neovascularization and tumor growth. J Clin Invest 115:2979–2991

Naldini A, Carraro F (2005) Role of inflammatory mediators in angiogenesis. Curr Drug Targets Inflamm Allergy 4:3–8

Naldini A, Leali D, Pucci A et al (2006) Cutting edge: IL-1beta mediates the proangiogenic activity of osteopontin-activated human monocytes. J Immunol 177:4267–4270

Narayana A, Kelly P, Golfinos J et al (2008) Antiangiogenic therapy using bevacizumab in recurrent high-grade glioma: impact on local control and patient survival. J Neurosurg 110:173–180

Nasarre P, Thomas M, Kruse K et al (2009) Host-derived angiopoietin-2 affects early stages of tumor development and vessel maturation but is dispensable for later stages of tumor growth. Cancer Res 69:1324–1333

Natale RB, Bodkin D, Govindan R et al (2009) Vandetanib versus gefitinib in patients with advanced non-small-cell lung cancer: results from a two-part, double-blind, randomized phase ii study. J Clin Oncol 27:2523–2529

Nemeth JA, Nakada MT, Trikha MA (2007) Alpha-v integrins as therapeutic targets in oncology. Cancer Invest 25:632–646

Niakosari F, Kahn HJ, McCready D et al (2008) Lymphatic invasion identified by monoclonal antibody D2-40, younger age, and ulceration: predictors of sentinel lymph node involvement in primary cutaneous melanoma. Arch Dermatol 144:462–467

Nico B, Crivellato E, Guidolin D et al (2010) Intussusceptive microvascular growth in human glioma. Clin Exp Med 10:93–98

Nicosia RF, Tuszynski GP (1994) Marix-bound thrombospondin promotes angiogenesis in vitro. J Cell Biol 124:183–193

Nielsen BS, Sehested M, Kjeldsen L et al (1997) Expression of matrix metalloprotease-9 in vascular pericytes in human breast cancer. Lab Invest 77:345–355

Nishie A, Ono M, Shono T et al (1999) Macrophage infiltration and heme oxygenase-1 expression correlate with angiogenesis in human gliomas. Clin Cancer Res 5:1107–1113

Nilsson G, Forsberg-Nilsson K, Xiang Z et al (1997) Human mast cells express functional TrkA and are a source of nerve growth factor. Eur J Immunol 27:2295–2301

Nissen LJ, Cao R, Hedlund EM et al (2007) Angiogenic factors FGF2 and PDGF-BB synergistically promote murine tumor neovascularization and metastasis. J Clin Invest 117:2766–2777

Niwa Y, Akamatsu H, Niwa H et al (2001) Correlation of tissue and plasma RANTES levels with disease course in patients with breast or cervical cancer. Clin Cancer Res 7:285–289

Noh YH, Matsuda K, Hong YK et al (2003) An N-terminal 80 kDa recombinant fragment of human thrombospondin-2 inhibits vascular endothelial growth factor induced endothelial cell migration in vitro and tumor growth and angiogenesis in vivo. J Invest Dermatol 121:1536–1543

Norden AD, Drappatz J, Wen PY (2009) Antiangiogenic therapies for high-grade glioma. Nat Rev Neurol 5:610–620

Norden-Zfoni A, Desai J, Manola J et al (2007) Blood-based biomarkers of SU11248 activity and clinical outcome in patients with metastatic imatinib-resistant gastrointestinal stromal tumor. Clin Cancer Res 13:2643–2650

Norrby K, Jakobsson A, Sörbo J (1986) Mast cell-mediated angiogenesis: a novel experimental model using the rat mesentery. Virchows Arch B Cell Pathol Mol Pathol 52:195–206

Norrby K, Jakobsson A, Sörbo J (1989) Mast-cell secretion and angiogenesis, a quantitative study in rats and mice. Virchows Arch B Cell Pathol Mol Pathol 57:251–256

Norden AD, Young GS, Setayesh K et al (2008) Bevacizumab for recurrent malignant gliomas: efficacy, toxicity and patterns of recurrence. Neurology 70:779–787

North T, Gut L, Stacy T et al (1999) Cbfa2 is required for the formation of intra-aortic hematopoietic clusters. Development 126:2563–2575

Northfelt DW, Dezube BJ, Thommes J (1997) Efficacy of pegylated-liposomal doxorubicin in the treatment of AIDS-related Kaposi's sarcoma after failure of standard chemotherapy. J Clin Oncol 15:653–659

Nozawa H, Chiu C, Hanahan D (2006) Infiltrating neutrophils mediate the initial angiogenic switch in a mouse model of multistage carcinogenesis. Proc Natl Acad Sci U S A 103:12493–12498

Nowak G, Karrar A, Holmen C et al (2004) Expression of vascular endothelial growth factor receptor-2 or Tie-2 on peripheral blood cells defines functionally competent cell population capable of reendothelization. Circulation 110:3699–3707

Nyberg P, Salo T, Kalluri R (2008) Tumor microenvironment and angiogenesis. Front Biosci 13:6537–6553

O'Connell PJ, Gerkis V, d'Apice AJ (1991) Variable O-glycosylation of CD13 (aminopeptidase N). J Biol Chem 266:4593–4597

Obermueller E, Vosseler S, Fusenig NE et al (2004) Cooperative autocrine and paracrine functions of granulocyte colony-stimulating factor and granulocyte-macrophage colony-stimulating factor in the progression of skin carcinoma cells. Cancer Res 64:7801–7812

Ohki Y, Heissig B, Sato Y et al (2005) Granulocyte colony-stimulating factor promotes neo-vascularization by releasing vascular endothelial growth factor from neutrophils. FASEB J 19:2005–2007

Ohno F, Nakanishi H, Abe A et al (2007) Regional difference in intratumoral lymphangiogenesis of oral squamous cell carcinoma evaluated by immunohistochemistry using D2-40 and podoplanin antibody: an analysis in comparison with angiogenesis. J Oral Pathol Med 36:281–289

Ohno S, Ohno Y, Suzuki N et al (2004) Correlation of histological localization of tumor-associated macrophages with clinicopathological features in endometrial cancer. Anticancer Res 24:3335–3342

Okamoto I, Kaneda H, Satoh T et al (2010) Phase I safety, pharmacokinetic, and biomarker study of BIBF 1120, an oral triple tyrosine kinase inhibitor in patients with advanced solid tumors. Mol Cancer Ther 9:2825–2833

Omachi T, Kawai Y, Mizuno R et al (2007) Immunohistochemical demonstration of proliferating lymphatic vessels in colorectal carcinoma and its clinicopathological significance. Cancer Lett 246:167–172

O'Reilly MS, Holmgren L, Shing Y et al (1994) Angiostatin: a novel angiogenesis inhibitor that mediates the suppression of metastases by a Lewis lung carcinoma. Cell 79:315–328

O'Reilly MS, Holmgren L, Chen C et al (1996) Angiostatin induces and sustains dormancy of human primary tumors in mice. Nature Med 2:689–692

O'Reilly MS, Boehm T, Shing Y et al (1997) Endostatin: an endogenous inhibitor of angiogenesis and tumor growth. Cell 88:277–285

Orlic D, Kajstura J, Chimenti S et al (2001) Mobilized bone marrow cells repair the infarcted heart, improving function and survival. Proc Natl Acad Sci U S A 98:10344–10349

Otsu K, Das S, Houser SD et al (2009) Concentration-dependent inhibition of angiogenesis by mesenchymal stem cells. Blood 113:4197–4205

Padera TP Stoll BR, Tooredman JB et al (2004) Pathology: cancer cells compress intratumoral vessels. Nature 427:695

Padera TP, Kuo AH, Hoshida T et al (2008) Differential response of primary tumor versus lymphatic metastasis to VEGFR-2 and VEGFR-3 kinase inhibitors cediranib and vandetanib. Mol. Cancer Res 7:2272–2279

Paez-Ribes M, Allen A, Hudock J et al (2009) Antiangiogenic therapy elicits malignant progression of tumors to increased local invasion and distant metastasis. Cancer Cell 15:222–231

Paku S, Deszo K, Bugyik E et al (2011) A new mechanism for pillar formation during tumor-induced intussusceptive angiogenesis. Inverse sprouting. Am J Pathol 179:1573–1585

Pan PY, Wang GX, Yin B et al (2008) Reversion of immune tolerance in advanced malignancy: modulation of myeloid-derived suppressor cell development by blockade of stem-cell factor function. Blood 111:219–228

Pang RW, Poon RT (2006) Clinical implications of angiogenesis in cancers. Vasc Health Risk Manag 2:97–108

Panka DJ, Wang W, Atkins MB et al (2006) The Raf inhibitor BAY 43–9006 (Sorafenib) induces caspase-independent apoptosis in melanoma cells. Cancer Res 66:1611–1619

Papetti M, Herman IM (2002) Mechanisms of normal and tumor-derived angiogenesis. Am J Physiol Cell Physiol 282:C947–C970

Parker BS, Argani P, Cook BP et al (2004) Alterations in vascular gene expression in invasive breast carcinoma. Cancer Res 64:7857–7866

Pasqualini R, Ruoslahti E (1996) Organ targeting in vivo using phage display peptide libraries. Nature 380:364–366

Passalidou E, Trivella M, Shing N et al (2002) Vascular phenotype in angiogenic and antiangiogenic lung non-small cell carcinomas. Br J Cancer 86:244–249

Pastorino F, Brignole C, Marimpietri D et al (2003a) Vascular damage and anti-angiogenic effects of tumor vessel-targeted liposomal chemotherapy. Cancer Res 63:7400–7409

Pastorino F, Brignole C, Marimpietri D et al (2003b) Doxorubicin-loaded Fab' fragments of anti-disialoganglioside immunoliposomes selectively inhibit the growth and dissemination of human neuroblastoma in nude mice. Cancer Res 2003 63:86–92

Pastorino F, Brignole C, Di Paolo D et al (2006) Targeting liposomal chemotherapy via both tumor cell-specific and tumor vasculature-specific ligands potentiates therapeutic efficacy. Cancer Res 66:10073–10082

Pastorino F, Marimpietri D, Brignole C et al (2007) Ligand-targeted liposomal therapies of neuroblastoma. Curr Med Chem 14:3070–3078

Pastorino F, Di Paolo D, Piccardi F (2008a) Enhanced antitumor efficacy of clinical-grade vasculature-targeted liposomal doxorubicin. Clin Cancer Res 14:7320–7329

Pastorino F, Mumbengegwi DR, Ribatti D et al (2008b) Increase of therapeutic effects by treating melanoma with targeted combinations of c-myc antisense and doxorubicin. J Control Release 126:85–94

Patan S, Munn LL, Jain RK (1996) Intussusceptive microvascular growth in a human colon adenocarcinoma xenograft: a novel mechanism of tumor angiogenesis. Microvasc Res 51:260–272

Peichev M, Naiyer AJ, Pereira D et al (2000) Expression of VEGFR-2 and AC133 by circulating human CD34(+) cells identifies a population of functional endothelial precursors. Blood 95:952–958

Pennacchietti S, Michieli P, Galluzzo M et al (2003) Hypoxia promotes invasive growth by transcriptional activation of the met protooncogene. Cancer Cell 3:347–361

Petras K, Hanahan D (2005) A multitargeted, metronomic, and maximum-tolerated dose "chemo switch" regimen is antiangiogenic, producing objective responses and survival benefit in a mouse model of cancer. J Clin Oncol 23:939–952

Pezzella F, Pastorino U, Tagliabue E et al (1997) Non-small-lung carcinoma tumor growth without morphological evidence of neo-angiogenesis. Am J Pathol 151:1417–1423

Pezzolo A, Parodi F, Corrias MV et al (2007) Tumor origin of endothelial cells in human neuroblastoma. J Clin Oncol 25:376–383

Pircher A, Kähler CM, Skvortsov S et al (2008) Increased numbers of endothelial progenitor cells in peripheral blood and tumor specimens in non small cell lung cancer: a methodological challenge and an ongoing debate on the clinical relevance. Oncol Rep 19:345–352

Plate KH, Breier G, Widch HA et al (1992) Vascular endothelial growth factor is a potent tumour angiogenesis factor in human gliomas *in vivo*. Nature 359:845–848

Potenta S, Zeisberg E, Kalluri R (2008) The role of endothelial-to-mesenchymal transition in cancer progression. Br J Cancer 99:1375–1379

Potgens AJ, Lubsen NH, van Altena MC et al (1995) Vascular permeability factor expression influences tumor angiogenesis in human melanoma lines xenografted to nude mice. Am J Pathol 146:197–209

Puxeddu I, Alian A, Piliponsky AM et al (2005) Human peripheral blood eosinophils induce angiogenesis. Int J Biochem Cell Biol 37:628–636

Puxeddu I, Berkman N, Nissim Ben Efraim AH et al (2009) The role of eosinophil major basic protein in angiogenesis. Allergy 64:368–374

Qian X, Wang TN, Rothman VL et al (1997) Thrombospondin-1 modulates angiogenesis in vitro by up-regulation of matrix metalloproteinase-9 in endothelial cells. Exp Cell Res 235:403–412

Qu Z, Liebler JM, Powers MR et al (1995) Mast cells are a major source of basic fibroblast growth factor in chronic inflammation and cutaneous hemangioma. Am J Pathol 147:564–573

Qu Z, Huang X, Ahmadi P et al (1998a) Synthesis of basic fibroblast growth factor by murine mast cells. Regulation by transforming growth factor beta, tumor necrosis factor alpha, and stem cell factor. Int Arch Allergy Immunol 115:47–54

Qu Z, Kayton RJ, Ahmadi P et al (1998b) Ultrastructural immunolocalization of basic fibroblast growth factor in mast cell secretory granules. Morphological evidence for bfgf release through degranulation. J Histochem Cytochem 46:1119–1128

Queen MM, Ryan RE, Holzer RG et al (2005) Breast cancer cells stimulate neutrophils to produce oncostatin M: potential implications for tumor progression. Cancer Res 65:8896–8904

Rafii S, Lyden D, Benezra R et al (2002) Vascular and haematopoietic stem cells: novel targets for anti-angiogenesis therapy? Nat Rev Cancer 2:826–835

Rafii S, Lyden D (2008) Cancer. A few to flip the angiogenic switch. Science 319:163–164

Raica M, Ribatti D (2010) Targeting tumor lymphangiogenesis: an update. Curr Med Chemistry 17:698–708

Raica M, Ribatti D, Mogoanta L et al (2008) Podoplanin expression in advanced-stage gastric carcinoma and prognostic value of lymphatic microvessel density. Neoplasma 55:454–459

Rak J, Mitsuhashi Y, Bayko L et al (1995) Mutant ras oncogenes upregulate VEGF/VPF expression: implications for induction and inhibition of tumor angiogenesis. Cancer Res 55:4575–4580

Ramalingam SS, Belani CP, Mack PC et al (2010) Phase II study of Cediranib (AZD 2171), an inhibitor of the vascular endothelial growth factor receptor, for second-line therapy of small cell lung cancer (National Cancer Institute #7097). J Thorac Oncol 5:1279–1284

Ratain MJ, Eisen T, Stadler WM et al (2006) Phase II placebo-controlled randomized discontinuation trial of sorafenib in patients with metastatic renal cell carcinoma. J Clin Oncol 24:2505–2512

Raza A, Franklin MJ, Dudek AZ (2010) Pericytes and vessel maturation during tumor angiogenesis and metastasis. Am J Hematol 85:593–598

Reck M, Kaiser R, Eschbach C et al (2011) A phase II double-blind study to investigate efficacy and safety of two doses of the triple angiokinase inhibitor BIBF 1120 in patients with relapsed advanced non-small cell lung cancer. Ann Oncol 22:1374–1381

Reed JA, McNutt NS, Bogdany JK et al (1986) Expression of the mast cell growth factor interleukin-3 in the melanocytic lesions correlates with an increased number of mast cells in the perilesional stroma: implications for melanoma progression. J Cutan Pathol 23:495–505

Rehman J, Li J, Orschell CM, March KL (2003) Peripheral blood "endothelial progenitor cells" are derived from monocyte/macrophages and secrete angiogenic growth factors. Circulation 107:1164–1169

Reinmuth N, Liu W, Jung YD et al (2001) Induction of VEGF in perivascular cells defines a potential paracrine mechanism for endothelial cell survival. FASEB J 15:1239–1241

Relf M, LeJeune S, Scott PA et al (1997) Expression of the angiogenic factors vascular endothelial cell growth factor, acidic and basic fibroblast growth factor, tumor growth factor beta-1, platelet-derived endothelial cell growth factor, placenta growth factor, and pleiotrophin in human primary breast cancer and its relation to angiogenesis. Cancer Res 57:963–969

Renyi-Vamos F, Tovari J, Fillinger J et al (2005) Lymphangiogenesis correlates with lymph node metastasis, prognosis, and angiogenic phenotype in human non-small cell lung cancer. Clin Cancer Res 11:7344–7353

Reyes M, Lund T, Lenvik T et al (2001) Purification and ex vivo expansion of postnatal human marrow mesodermal progenitor cells. Blood 98:2615–2625

Ria R, Todoerti K, Berardi S et al (2009) Gene expression profiling of bone marrow endothelial cells in patients with multiple myeloma. Clin Cancer Res 15:5369–5378

Ribatti D (2010) Biomarkers of response to angiogenesis inhibitors: an open and unsolved question. Eur J Cancer 46:6–8

Ribatti D (2011) Antiangiogenic therapy accelerates tumor metastasis. Leuk Res 35:24–26

Ribatti D, Crivellato E (2009) Immune cells and angiogenesis. J Cell Mol Med 13:2822–2833

Ribatti D, Vacca A (2005) Therapeutic renaissance of thalidomide in the treatment of haematological malignancies. Leukemia 19:1523–1531

Ribatti D, Vacca A (2008) Overview of tumor angiogenesis. In: Figg WD, Folkman J (eds) Angiogenesis. An integrative approach from science to medicine. Springer, New York, pp 161–168

Ribatti D, Roncali L, Nico B et al (1987) Effects of exogenous heparin on the vasculogenesis of the chorioallantoic membrane. Acta Anat 130:257–263

Ribatti D, Vacca A, Nico B et al (1996) Angiogenesis spectrum in the stroma of B-cell non Hodgkin's lymphomas. An immunohistochemical and ultrastructural study. Eur J Haematol 56:45–53

Ribatti D, Nico B, Vacca A et al (1998) Do mast cells help to induce angiogenesis in B-cell non-Hodgkin's lymphomas? Br J Cancer 77:1900–1906

Ribatti D, Vacca A, Nico B et al (1999a) Bone marrow angiogenesis and mast cell density increase simultaneously with progression of human multiple myeloma. Br J Cancer 79:451–455

Ribatti D, Vacca A, Dammacco F (1999b) The role of the vascular phase in solid tumor growth: a historical review. Neoplasia 1:293–302

Ribatti D, Vacca A, Marzullo A et al (2000) Angiogenesis and mast cell density with tryptase activity increase simultaneously with pathological progression in B-cell non-Hodgkin's lymphomas. Int J Cancer 82:171–175

Ribatti D, Crivellato E, Candussio L et al (2001) Mast cells and their secretory granules are angiogenic in the chick embryo chorioallantoic membrane. Clin Exp Allergy 31:602–608

Ribatti D, Polimero G, Vacca A et al (2002) Correlation of bone marrow angiogenesis and mast cells with tryptase activity in myelodysplastic syndromes. Leukemia 16:1680–1684

Ribatti D, Vacca A, Ria R et al (2003a) Neovascularization, expression of fibroblast growth factor-2, and mast cells with tryptase activity increase simultaneously with pathological progression in human malignant melanoma. Eur J Cancer 39:666–674

Ribatti D, Ennas MG, Vacca A et al (2003b) Tumor vascularity and tryptase-positive mast cells correlate with a poor prognosis in melanoma. Eur J Clin Invest 33:420–425

Ribatti D, Molica S, Vacca A et al (2003c) Tryptase-positive mast cells correlate positively with bone marrow angiogenesis in B-cell chronic lymphocytic leukemia. Leukemia 17:1428–1430

Ribatti D, Nico B, Floris C et al (2005a) Microvascular density, vascular endothelial growth factor immunoreactivity in tumor cells, vessel diameter and intussusceptive microvascular growth in primary melanoma. Oncol Rep 14:81–84

Ribatti D, Finato N, Crivellato E et al (2005b) Neovascularization and mast cells with tryptase activity increase simultaneously with pathological progression in human endometrial cancer. Am J Obstet Gynecol 193:1961–1965

Ribatti D, Nico B, Vacca A (2006) Importance of the bone marrow microenvironment in inducing the angiogenic response in multiple myeloma. Oncogene 25:4257–4266

Ribatti D, Nico B, Crivellato E et al (2007a) The history of the angiogenic switch concept. Leukemia 21:44–52

Ribatti D, Nico B, Crivellato E et al (2007b) Macrophages and tumor angiogenesis. Leukemia 21:2085–2089

Ribatti D, Nico B, Crivellato E et al (2007c) The structure of the vascular network of tumors. Cancer Letters 248:18–23

Ribatti D, Guidolin D, Marzullo A et al (2010) Mast cells and angiogenesis in gastric carcinoma. Int J Exp Pathol 91:350–356

Ribatti D, Nico B, Crivellato E (2011a) The role of pericytes in angiogenesis. Int J Dev Biol 55:261–268

Ribatti D, Ranieri G, Nico B et al (2011b) Tryptase and chymase are angiogenic *in vivo* in the chorioallantoic membrane assay. Int J Dev Biol 55:99–102

Ribatti D, Djonov V (2012) Intussusceptive microvascular growth in tumors. Cancer Lett 316:126–131

Riboldi E, Musso T, Moroni E et al (2005) Cuting edge: proangiogenic properties of alternatively activated dendritic cells. J Immunol 175:2788–2792

Ricci-Vitiani L, Pallini R, Biffoni M et al (2010) Tumor vascularization via endothelial differentiation of glioblastoma stem-like cells. Nature 468:824–828

Rigolin GM, Fraulini C, Ciccone M et al (2006) Neoplastic circulating endothelial cells in multiple myeloma with 13q14 deletion. Blood 107:2531–2535

Rini BI, Garcia J, Cooney MM et al (2009) A phase I study of sunitinib plus bevacizumab in advanced solid tumours. Clin Cancer Res 5:6227

Rolny C, Mazzone M, Tugnes S et al (2011) HRG inhibits tumor growth and metastasis by inducing macrophage polarization and vessel normalization through downregulation of PlGF. Cancer Cell 19:31–44

Romagnani P, Annunziato F, Lasagni L et al (2001) Cell cycle-dependent expression of CXC chemokine receptor 3 by endothelial cells mediates angiostatic activity. J Clin Invest 107:53–63

Romagnani P, Annunziato F, Liotta F et al (2005) CD14$^+$ CD34 low cells with stem cell phenotypic and functional features are the major source of circulating endothelial progenitors. Circ Res 97:314–322

Roskoski R Jr (2007) Sunitinib: a VEGF and PDGF receptor protein kinase and angiogenesis inhibitor. Biochem Biophys Res Commun 356:323–328

Rowand JL, Martin G, Doyle GV et al (2007) Endothelial cells in peripheral blood of helathy subjects and patients with metastatic carcinomas. Cytometry A 71:105–113

Saharinien P, Tammela T, Karkakainen MJ et al (2004) Lymphatic vasculature: development, molecular regulation and role in tumor metastasis and inflammation. Trends Immunol 25:387–395

Sahni D, Robson A, Orchard G et al (2005) The use of LYVE-1 antibody for detecting lymphatic involvement in patients with malignant melanoma of known sentinel node status. J Clin Pathol 58:715–721

Saidi A, Hagedorn M, Allain N et al (2009) Combined targeting of interleukin 6 and vascular endothelial growth factor potently inhibits glioma growth and invasiveness. Int J Cancer 125:1054–1064

Salven P, Mustjoki S, Alitalo R et al (2003) VEGFR-3 and CD133 identify a population of CD34$^+$ lymphatic/vascular endothelial precursor cells. Blood 101:168–172

Sandler A, Gray R, Perry MC et al (2006) Paclitaxel-carboplatin alone or with bevacizumab for non-small-cell lung cancer. N Engl J Med 355:2542–2550

Santamaria M, Moscatelli G, Viale GL et al (2003) Immunoscintigraphic detection of the ED-B domain of fibronectin, a marker of angiogenesis, in patients with cancer. Clin Cancer Res 9:571–579

Santerelli JG, Udani V, Yung YC et al (2006) Incorporation of bone marrow-derived Flk-1-expressing CD34$^+$ cells in the endothelium of tumour vessels in the mouse brain. Neurosurgery 59:374–382

Sapra P, Allen TM (2002) Internalizing antibodies are necessary for improved therapeutic efficacy of antibody-targeted liposomal drugs. Cancer Res 62:7190–194

Satchi-Fainaro R, Mamluk R, Wang L et al (2005) Inhibition of vessel permeability by TNP-470 and its polymer conjugate, caplostatin. Cancer Cell 7:251–261

Sato M, Arap W, Pasqualini R (2007) Molecular targets on blood vessels for cancer therapies in clinical trials. Oncology (Williston Park) 21:1346–1352

Sato TN, Quin Y, Kozak CA et al (1993) Tie-1 and Tie-2 define another class of putative receptor tyrosine kinase genes expressed in early embryonic vascular system. Proc Natl Acad Sci U S A 90:9355–9358

Sato Y (2011) Persistent vascular normalization as an alternative goal of anti-angiogenic cancer therapy. Cancer Sci. 102:1253–1256 (Epub ahead of print)

Sawatsubashi M, Yamada T, Fukushima N et al (2000) Association of vascular endothelial growth factor and mast cells with angiogenesis in laryngeal squamous cell carcinoma. Virchows Arch B Cell Pathol Mol Pathol 436:243–248

Scapini P, Nesi L, Morini M et al (2002) Generation of biologically active angiostatin kringle 1–3 by activated human neutrophils. J Immunol 168:5798–5804

Schacht V, Dadras SS, Johnson LA et al (2005) Up-regulation of the lymphatic marker podoplanin, a mucin-type glycoprotein, in human squamous cell carcinoma and germ cell tumors. Am J Pathol 166:913–921

Schatteman GC, Hanlon HD, Jiao G et al (2000) Blood-derived angioblasts accelerate blood-flow restoration in diabetic mice. J Clin Invest 106:571–578

Schlingemann RO, Rietveld FJ, De Wall RM et al (1990) Expression of the high molecular weight melanoma-associated antigen by pericytes during angiogenesis in tumors and in healing wounds. Am J Pathol 136:1393–1405

Schmeisser A, Garlichs CD, Zhang H et al (2001) Monocytes coexpress endothelial and macrophagocytic lineage markers and form cord-like structures in Matrigel under angiogenic conditions. Cardiovasc Res 49:671–680

Schmid MC, Varner JA (2007) Myeloid cell trafficking and tumor angiogenesis. Cancer Lett 250:1–8

Schmidt CR, Panageas KS, Coit S et al (2009) An increased number of sentinel lymph nodes is associated with advanced Breslow depth and lymphovascular invasion in patients with primary melanoma. Ann Surg Oncol 16:948–952

Schomber T, Zumsteg A, Strittmatter K et al (2009). Differential effects of the vascular endothelial growth factor receptor inihibitor PTK787/ZK222584 on tumor angiogenesis and lymphangiogenesis. Mol Cancer Ther 8:55–63

Schoppmann SF, Mayer G, Aumayr K et al (2004) Austrian breast and colorectal cancer study group. Prognostic value of lymphangiogenesis and lymphovascular invasion in breast cancer. Ann Surg 240:306–312

Schoppmann SF, Fenzl A, Schindl M et al (2006) Hypoxia inducible factor-1α correlates with VEGF-C expression and lymphangiogenesis in breast cancer. Breast Cancer Res. Treat 99:135–141

Schnurch H, Risau W (1993) Expression of Tie-2, a member of a novel family of receptor tyrosine kinases, in the endothelial cell lineage. Development 119:957–968

Schruefer R, Sulyok S, Schymeinsky J et al (2006) The proangiogenic capacity of polymorphonuclear neutrophils delineated by microarray technique and by measurement of neovascularization in wounded skin of CD18-deficient mice. J Vasc Res 43:1–11

Schwartz JD, Monea S, Marcus SG et al (1998) Soluble factor(s) released from neutrophils activates endothelial cell matrix metalloproteinase-2. J Surg Res 76:79–85

Seaman S, Stevens J, Yang MY et al (2007) Genes that distinguish physiological and pathological angiogenesis. Cancer Cell 11:539–554

Seandel M, Butler J, Lyden D et al (2008) A catalytic role for proangiogenic marrow-derived cells in tumor neovascularization. Cancer Cell 13:181–183

Seftor RE, Seftor EA, Koshikawa N et al (2001) Cooperative interactions of laminin 5 gamma-2 chain, matrix metalloproteinase-2, and membrane type-1-matrix/metalloproteinase are required for mimicry of embryonic vasculogenesis by aggressive melanoma. Cancer Res 61:6322–6327

Seiwert TY, Haraf DJ, Cohen EE et al (2008) Phase I study of bevacizumab added to fluorouracil- and hydroxyurea-based concomitant chemoradiotherapy for poor-prognosis head and neck cancer. J Clin Oncol 26:1732–1741

Semenza GL (1996) Transcriptional regulation by hypoxia-inducible factor-1. Trends Cardiovasc Med 6:151–157

Semenza GL (2003) Targeting HIF-1 for cancer therapy. Nat Rev Cancer 3:721–732

Sennino B, Falcon BL, Mccauley D et al (2007) Sequential loss of tumor vessel pericytes and endothelial cells after inhibition of platelet-derived growth factor B by selective aptamer AX102. Cancer Res 67:7358–7367

Sergeeva A, Kolonin MG, Molldrem J et al (2006) Display technologies: application for the discovery of drug and gene delivery agents. Adv Drug Deliv Rev 58:1622–1654

Shaheen RM, Tseng WW, Davis DW et al (2001) Tyrosine kinase inhibition of multiple angiogenic growth factor receptors improves survival in mice bearing colon cancer liver metastases by inhibition of endothelial cell survival mechanism. Cancer Res 61:1464–1468

Shaked Y, Bertolini F, Man S et al (2005) Genetic heterogeneity of the vasculogenic phenotype parallels angiogenesis: implications for cellular surrogate marker analysis of antiangiogenesis. Cancer Cell 27:101–111

Shaked Y, Ciarrocchi A, Franco M et al (2006) Therapy-induced acute recruitment of circulating endothelial progenitor cells in tumors. Science 313:1785–1787

Shamamian P, Schwartz JD, Pocock BJ et al (2001) Activation of progelatinase A (MMP-2) by neutrophil elastase, cathepsin G, and proteinase-3: a role for inflammatory cells in tumor invasion and angiogenesis. J Cell Physiol 189:197–206

Sharma A, Trivedi NR, Zimmerman MA et al (2005) Mutant V599EB-Raf regulates growth and vascular development of malignant melanoma tumors. Cancer Res 65:2412–2421

Sharma N, Seftor RE, Seftor EA et al (2002) Prostatic tumor cell plasticity involves cooperative interactions of distinct phenotypic subpopulations: role in vasculogenic mimicry. Prostate 50:189–201

Shields JD, Borsetti M, Rigby H et al (2004) Lymphatic density and metastatic spread in human malignant melanoma. Br J Cancer 90:693–700

Shirakawa K, Kobayashi H, Heike Y et al (2002) Hemodynamics in vasculogenic mimicry and angiogenesis of inflammatory breast cancer xenograft. Cancer Res 62:560–566

Shojaei F, Ferrara N (2007) Antiangiogenic therapy for cancer: an update. Cancer J 13:345–348

Shojaei F, Ferrara N (2008) Role of microenvironment in tumor growth and in refractoriness/resistance to antiangiogenic therapies. Drug Resist Update 11:219–230

Shojaei F, Wu X, Qu M et al (2009) G-CSF-initiated myeloid cell mobilization and angiogenesis mediate tumor refractoriness to anti-VEGF therapy in mouse models. Proc Natl Acad Sci U S A 106:6742–6747

Shojaei X, Wu AK, Malik C et al (2007) Tumor refractorriness to anti-VEGF treatment is mediated by CD11$^+$ Gr1$^+$ myeloid cells. Nat Biotechnol 25:911–920

Shweiki D, Itin A, Soffer D et al (1992) Vascular endothelial growth factor induced by hypoxia may mediate hypoxia-initiated angiogenesis. Nature 359:843–845

Sica A, Schioppa T, Mantovani A et al (2006) Tumour-associated macrophages are a distinct M2 polarised population promoting tumour progression: potential targets of anti-cancer therapy. Eur J Cancer 42:717–727

Sidky YA, Borden EC (1987) Inhibition of angiogenesis by interferons: effects on tumor- and lymphocyte-induced vascular responses. Cancer Res 47:5155–5161

Sie M, Wagemakers M, Molem G et al (2009) The angiopoietin-1/angiopoietin-2 balance as a prognostic marker in primary glioblastoma multiforme. J Neurosurg 110:147–155

Sipos B, Klapper W, Kruse ML et al (2004) Expression of lymphangiogenic factors and evidence of intratumoral lymphangiogenesis in pancreatic endocrine tumors. Am J Pathol 165:1187–1197

Skobe M, Hawighorts T, Jackson DG et al (2001) Induction of tumor lymphangiogenesis by VEGF-C promotes breast cancer metastasis. Nature Med 7:192–198

Smith JK, Mamoon NM, Duher RJ (2004) Emerging roles of targeted small molecule protein-tyrosine kinase inhibitors in cancer therapy. Oncol Res 14:175–225

Smith NR, James NH, Oakley I et al (2007) Acute pharmacodynamic and antivascular effects of the vascular endothelial growth factor signaling inhibitor AZD2171 in Calu-6 human lung tumor xenografts. Mol Cancer Ther 6:2198–2208

Solomon A, Aloe L, Pe'er J et al (1998) Nerve growth factor is preformed in and activates human peripheral blood eosinophils. J Allergy Clin Immunol 102:454–460

Solovey AN, Gui L, Chang L et al (2001) Identification and functional assessment of endothelial P1H12. J Lab Clin Med 138:322–331

Song S, Wientjes MG, Walsh C, Au JL (2001) Nontoxic doses of suramin enhance activity of paclitaxel against lung metastases. Cancer Res 61:6145–6150

Song S, Ewald AJ, Stallcup W et al (2005) PDGFRbeta$^+$ prerivascular progenitor cells in tumours regulate pericyte differentiation and vascular survival. Nat Cell Biol 7:870–879

Sood AK, Fletcher MS, Zahn CM et al (2002) The clinical significance of tumor cell-lined vasculature in ovarian carcinoma: impications for anti-vasculogenic therapy. Cancer Biol Ther 1:661–664

Sörbo J, Jakobbson A, Norrby K (1994) Mast cell histamine is angiogenic through receptors for histamine 1 and histamine 2. Int J Exp Pathol 75:43–50

Sorensen AG, Batchelor TC, Zhang WT et al (2009) A "vascular normalization index" as a potential mechanistic biomarker to predict survival after a single dose od cediranib in recurrent glioblastoma patients. Cancer Res 69:5296–5300

Sosman JA, Puzanov I, Atkins MB (2007) Opportunities and obstacles to combination targeted therapy in renal cell cancer. Clin Cancer Res 13:764s–769s

Sosman JA, Flaherty KT, Atkins MB et al (2008) Updated results of phase I trial of sorafenib (S) and bevacizumab (B) in patients with metastatic renal cell cancer (mRCC). J Clin Oncol 26:5011 (Abstract)

Soucek L, Lawlar ER, Soto D et al (2007) Mast cells are required for angiogenesis and macroscopic expansion of Myc-induced pancreatic islet tumors. Nat Med 13:1211–1218

Spurbeck WW, Ng CY, Strom TS et al (2002) Enforced expression of tissue inhibitor of matrix metalloproteinase-3 affects functional capillary morphogenesis and inhibits tumor growth in a murine tumor model. Blood 100:3361–3368

St Croix B, Rago C, Velculescu V et al (2000) Gene expressed in human tumor and endothelium. Science 289:1197–1202

Stacker SA, Caesar C, Baldwin ME et al (2001) VEGF-D promotes the metastatic spread of tumor cells via the lymphatics. Nat Med 7:186–191

Stacker SA, Achen MG, Jusssila L et al (2002) Lymphangiogenesis and cancer metastasis. Nature Rev Cancer 2:573–583

Starkey JR, Crowle PK, Taubenberger S (1988) Mast cell-deficient W/Wv mice exhibit a decreased rate of tumor angiogenesis. Int J Cancer 42:48–52

Stopfer P, Rathgen K, Bischoff D et al (2011) Pharmacokinetics and metabolism of BIBF 1120 after oral dosing to healthy male volunteers. Xenobiotica 41:297–311

Strander H (1986) Interferon treatment of human neoplasia. Adv Cancer Res 46:1–265

Strasly M, Cavallo F, Guena M et al (2001) IL-12 inhibition of endothelial cell functions and angiogenesis depends on lymphocyte-endothelial cell cross-talk. J Immunol 166:3890–3899

Sterinberg F, Rohrborn HJ, Otto T et al (1997) NIR reflection measurements of hemoglobin and cytochrome da3 in healthy tissue and tumors. Correlation to oxygen consuption: preclinical and clinical data. Adv Exp Med Biol 428:69–77

Strehlow K, Werner N, Berweiler J et al (2003) Estrogen increases bone marrow-derived endothelial progenitor cell production and diminishes neointima formation. Circulation 107:3059–3565

Streubel B. Chott A, Huber D et al (2004) Lymphoma-specific genetic aberrations in microvascular endothelial cells in B-cell lymphomas. N Engl J Med 351:250–259

Suda T, Takakura N, Oike Y (2000) Hematopoiesis and angiogenesis. Int J Hematol 71:99–107

Sugiura T, Inoue Y, Matsuki R et al (2009) VEGF-C and VEGF-D expression is correlated with lymphatic vessel density and lymph node metastasis in oral squamous cell carcinoma: implications for use as a prognostic marker. Intern J Oncol 34:673–680

Sullivan R, Graham CH (2007) Hypoxia-driven selection of the metastatic phenotype. Cancer Met Rev 26:319–331

Sun J, Wang Y, Chen Z et al (2009) Detection of lymphangiogenesis in non-small cell lung cancer and its prognostic value. J Exp Clin Cancer Res 28:21

Sund M, Hamano Y, Sugimoto H et al (2005) Function of endogenous inhibitors of angiogenesis as endothelium-specific tumor suppressors. Proc Natl Acad Sci U S A 102:2934–2939

Sunderkotter C, Goebeler M, Schiltze-Osthoff K et al (1991) Macrophage-derived angiogenesis factors. Pharmacol Ther 51:195–216

Suzuki E, Kapoor V, Jassar AS et al (2005) Gemcitabile selectively eliminates splenic Gr-1^+/CD11b$^+$ myeloid suppressor cells in tumor-bearing animals and enhances antitumor immune activity. Clin Cancer Res 11:6713–6721

Suzuki-Inoue K, Kato Y, Inoue O et al (2007) Involvement of the snake toxin receptor CLEC-2, in podoplanin-mediated platelet activation, by cancer cells. J Biol Chem 282:25993-26001

Szmit S, Zagrodzka M, Kurzyna M et al (2009) Sunitinib malate, a receptor tyrosine kinase inhibitor, is effective in the treatment of restrictive heart failure due to heart metastases from renal cell carcinoma. Cardiology 114:67–71

Taatjes DJ, Koch TH (2001) Nuclear targeting and retention of anthracycline antitumor drugs in sensitive and resistant tumor cells. Curr Med Chem 8:15–29

Takahashi T, Kalka C, Masuda H et al (1999) Ischemia- and cytokine-induced mobilization of bone marrow-derived endothelial progenitor cells for neovascularization. Nature Med 5:434–438

Takakura N, Huang XL, Naruse T et al (1998) Critical role of the TIE2 endothelial cell receptor in the development of definitive hematopoiesis. Immunity 9:677–686

Takeda M, Arao T, Yokote H et al (2007) AZD2171 shows potent antitumor activity against gastric cancer over-expressing fibroblast growth factor receptor 2/keratinocyte growth factor receptor. Clin Cancer Res 13:3051–3057

Takanami I, Takeuchi K, Naruke M (2000) Mast cell density is associated with angiogenesis and poor prognosis in pulmonary adenocarcinoma. Cancer 88:2686–2692

Tamura T, Minami H, Yamada Y et al (2006) A phase I dose-escalation study of ZD6474 in Japanese patients with solid, malignant tumors. J Thorac Oncol 1:1002–1009

Tang Y, Kim M, Carrasco D et al (2005) In vivo assessment of RAS-dependent maintenance of tumor angiogenesis by real-time magnetic resonance imaging. Cancer Res 65:8324–8330

Tannock IF (1968) The relation between cell proliferation and the vascular system in a transparent mouse mammary tumour. Br J Cancer 22:258–273

Taraboletti G, Roberts D, Liotta LA et al (1990) Platelet thrombospondin modulates endothelial cell adhesion, motility, and growth: a potential angiogenesis regulatory factor. J Cell Biol 111:765–772

Taraboletti G, Morbidelli L, Donnini S et al (2000) The heparin binding 25 kDa fragment of thrombospondin-1 promotes angiogenesis and modulates gelatinase and TIMP-2 production in endothelial cells. FASEB J 14:1674–1676

Taylor S, Folkman J (1982) Protamine is an inhibitor of angiogenesis. Nature 297:307–312

Tee MK, Vigne JL, Taylor RN (2006) All-trans retinoic acid inhibits vascular endothelial growth factor expression in a cell model of neutrophil activation. Endocrinology 147:1264–1270

Thelen A, Jonas S, Benckert C et al (2009) Tumor-associated lymphangiogenesis correlates with prognosis after resection of human hepatocellular carcinoma. Ann Surg Oncol 16:1222–1231

Theoharides TC, Conti P (2004) Mast cells: the Jekyll and Hyde of tumor growth. Trends Immunol 25:235–241

Thorpe PE (2004) Vascular targeting agents as cancer therapeutics. Clin Cancer Res 10:415–427

Tomita M, Matsuzaki Y, Onitsuka T (2000) Effect of mast cells on tumor angiogenesis in lung cancer. Ann Thorac Surg 69:1686–1689

Tong RT, Boucher Y, Kozin SV et al (2004) Vascular normalization by vascular endothelial growth factor receptor 2 blockade induces a pressure gradient across the vasculature and improves drug penetration in tumors. Cancer Res 64:3731–3736

Torchilin VP (2005) Recent advances with liposomes as pharmaceutical carriers. Nat Rev Drug Discov 4:145–160

Tóth-Jakatics R, Jimi S, Takebayashi S et al (2000) Cutaneous malignant melanoma: correlation between neovascularization and peritumoral accumulation of mast cells overexpressing vascular endothelial growth factor. Human Pathol 31:955–960

Tozer GM, Kanthou C, Baguley BC (2005) Disrupting tumour blood vessels. Nat Rev Cancer 5:423–435

Trinchieri G (1993) Interleukin-12 and its role in the generation of TH1 cells. Immunol Today 14:335–338

Turner HE, Nagy Z, Gatter KC et al (2000) Angiogenesis in pituitary adenomas and in the normal pituitary gland. J Clin Endocrinol Metab 85:1159–1162

Twardowski PW, Smith-Powell L, Carroll M et al (2008) Biologic markers of angiogenesis: circulating endothelial cells in patients with advanced malignancies treated on phase I protocol with metronomic chemotherapy and celecoxib. Cancer Invest 26:53–59

Ueno T, Toi M, Saji H et al (2000) Significance of macrophage chemottractant protein-1 in macrophage recruitment, angiogenesis and survival in human breast cancer. Clin Cancer Res 6:3282–3289

Urbich C, Heeschen C, Aicher A et al (2003) Relevance of monocytic features for neovascularization capacity of circulating endothelial progenitor cells. Circulation 108:2511–2516

Urbich C, Dimmeler S (2004) Endothelial progenitor cells functional characterization. Trends Cardiovasc Med 14:318–322

Uutela M, Wirzenius M, Paavonen K et al (2004) PDGF-D induces macrophage recruitment, increased interstitial pressure, and blood vessel maturation during angiogenesis. Blood 104:3198–3204

Vacca A, Ribatti D (2006) Bone marrow angiogenesis in multiple myeloma. Leukemia 20:193–199

Vacca A, Scavelli C, Montefusco V et al (2005) Thalidomide downregulates angiogenic genes in bone marrow endothelial cells of patients with active multiple myeloma. J Clin Oncol 25:5334–5346

Vajkoczy P, Farhadi M, Gaumann A et al (2002) Microtumor growth initiates angiogenic sprouting with simultanenous expression of VEGF, VEGF receptor-2, and angiopoietin-2. J Clin Invest 109:777–785

Van de Veire S, Stalmans I, Heindryckx F et al (2010) Further pharmacological and genetic evidence for the efficacy of PlGF inhibition in cancer and eye disease. Cell 141:178–190

Van den Eynden GG, Vandenberghe MK, van Dam PJH et al (2007) Increased sentinel lymph node lymphangiogenesis is associated with nonsentinel axillary lymph node involvement in breast cancer patients with a positive sentibel node. Clin Cancer Res 13:5391–5397

Van der Auwera I, van der Eynden GG, Colpaert CG et al (2005) Tumor lymphangiogenesis in inflammatory breast carcinoma: a histomorphometric study. Clin Cancer Res 11:7637–7642

Van der Schaft DW, Hillen F, Pawwels P et al (2005) Tumor cell plasticity in Ewing sarcoma, an alternative circulatory system stimulated by hypoxia. Cancer Res 65:11520–11528

Vasa M, Fitchtlscherer S, Aicher A et al (2001) Number and migratory activity of circulating progenitor cells inversely correlate with risk factors for coronary artery disease. Circ Res 89: E1–E7

Veeravagu A, Bababeyygy SR, Kalani MYS et al (2008) The cancer stem cell—vascular niche in brain tumor formation. Stem Cells Dev 17:859–868

Venneri MA, De Palma M, Ponzoni M et al (2007) Identification of proangiogenic TIE2-expressing monocytes (TEMs) in human peripheral blood and cancer. Blood 109:5276–5285

Vermeulen PB, Verhoeven D, Hubens G et al (1995) Microvessels density, endothelial cell proliferation and tumor cell proliferation in human colorectal adenocarcinomas. Ann Oncol 6:59–64

Vittet D, Grandini MH, Berthier R et al (1996) Embryonic stem cells differentiate in vitro to endothelial cells through successive maturation steps. Blood 88:3424–3431

Vu TH, Werb Z (2000) Matrix metalloproteinases: effectors of development and normal physiology. Genes Dev 14:2123–2133

Vleugel MM, Bos R, van der Groep P et al (2004) Lack of lymphangiogenesis during breast carcinogenesis. J Clin Pathol 57:746–751

Volker HU, Scheich M, Nowack I et al (2009) Lymphangiosis carcinomatosa in squamous cell carcinomas of larynx and hypopharynx—value of conventional evaluation and additional immunohistochemical staining of D2-40. World J Surg Oncol 7:25

Walsh LJ, Trinchieri G, Waldorf HA et al (1991) Human dermal mast cells contain and release tumor necrosis factor alpha, which induces endothelial leukocyte adhesion molecule 1. Proc Natl Acad Sci U S A 88:4220–4224

Watnick RS, Cheng YN, Rangarajan A et al (2003) Ras modulates Myc activity to repress thrombospondin-1 expression and increase tumor angiogenesis. Cancer Cell 3:219–231

Wesseling P, Schlingemann RO, Rietveld FJ et al (1995) Early and extensive contribution of pericytes/vascular smooth muscle cells to microvascular proliferation in glioblastoma multiforme: an immuno-light and immuno-electron microscopic study. J Neuropathol Exp Neurol 54:304–310

Whitehurst B, Flister MJ, Bagaitkar J et al (2007) Anti-VEGF-A therapy reduces lymphatic vessel density and expression of VEGFR-3 in an orthotopic breast tumor model. Int J Cancer 121:2181–2191

Wedge SR, Kendrew J, Hennequin LF et al (2005) AZD2171: a highly potent, orally bioavailable, vascular endothelial growth factor receptor-2 tyrosine kinase inhibitor for the treatment of cancer. Cancer Res 65:4389–4400

Wicki A, Christofori G (2007) The potential role of podoplanin in tumour invasion. Br J Cancer 96:1–5

Wicki A, Lehembre F, Wick N et al (2006) Tumor invasion in the absence of epithelial-mesenchymal transition: podoplanin-mediated remodeling of the actin cytoskeleton. Cancer Cell 9:261–272

Wierzbowska A, Robak T, Krawczynska A et al (2005) Circulating endothelial cells in patients with acute myeloid leukemia. Eur J Haematol 75:492–497

Wilhelm SM, Carter C, Tang L et al (2004) BAY 43-9006 exhibits broad spectrum oral antitumor activity and targets the RAF/MEK/ERK pathway and receptor tyrosine kinases involved in tumor progression and angiogenesis. Cancer Res 64:7099–7109

Wilhelm SM, Adnane L, Newell P et al (2008) Preclinical overview of sorafenib, a multikinase inhibitor that targets both Raf and VEGF and PDGF receptor tyrosine kinase signaling. Mol Cancer Ther 7:3129–3140

Willett CG, Boucher Y, di Tomaso E et al (2004) Direct evidence that the VEGF-specific antibody bevacizumab has antivascular effects in human rectal cancer. Nat Med 10:145–147

Willett CG, Boucher Y, Duda DG et al (2005) Surrogate markers for antiangiogenic therapy and dose-limiting toxicities for bevacizumab with radiation and chemoherapy: continued experience of a phase I trail in rectal cancer patients. J Clin Oncol 23:8136–8139

Williams CSM, Leek RD, Robson AM et al (2003) Absence of lymphangiogenesis and intratumoural lymph vessels in human metastatic breast cancer. J Pathol 200:195–206

Winkler F, Kozin SV, Tong RT et al (2004) Kinetics of vascular normalization by VEGFR2 blockade governs brain tumor response to radiation: role of oxygenation, angiopoietin-1 and matrix metalloproteinases. Cancer Cell 6:553–563

Wobster M, Siedel C, Schrama D et al (2006) Expression pattern of the lymphatic and vascular markers VEGFR-3 and CD31 does not predict regional lymph node metastasis in cutaneous melanoma. Arch Dermatol Res 297:352–357

Wong DT, Weller PF, Galli SJ et al (1990) Human eosinophils express transforming growth factor alpha. J Exp Med 172:673–681

Yamaguchi J, Kusano KF, Masuo O et al (2003) Stromal cell-derived factor-1 effects on ex vivo expanded endothelial progenitor cell recruitment for ischemic neovascularization. Circulation 107:1316–1322

Yang JC, Haworth L, Sherry RM et al (2003) A randomized trial of bevacizumab, an anti-vascular endothelial growth factor antibody for metastatic renal cancer. N Engl J Med 349:427–434

Yang L, DeBusk LM, Fukuda K et al (2004) Expansion of myeloid immune suppressor Gr$^+$ CD11b$^+$ cells in tumor-bearing host directly promotes tumor angiogenesis. Cancer Cell 6:409–421

Yang Q, Rasmussen SA, Friedman JM (2002) Mortality associated with Down's syndrome in the USA from 1983 to 1997: a population based study. Lancet 359:1019–1025

Yang Z, Poon RT (2008) Vascular changes in hepatocellular carcinoma. Anat Rec 291:721–734

Yao L, Sgadari C, Furuke K et al (1999) Contribution of natural killer cells to inhibition of angiogenesis by interleukin-12. Blood 93:1612–1621

Yao XH, Ping YF, Bian XW (2011) Contribution of cancer stem cells to tumor vasculogenic mimicry. Protein cell 2:266–272

Yin AH, Miraglia S, Zanjani ED et al (1997) AC133, a novel marker for human hematopoietic stem and progenitor cells. Blood 90:5002–5012

Yousefi S, Hemmann S, Weber M et al (1995) IL-8 is expressed by human peripheral blood eosinophils. Evidence for increased secretion in asthma. J Immunol 154:5481–5490

Yu C, Bruzek LM, Meng XW et al (2005) The role of Mcl-1 downregulation in the proapoptotic activity of the multikinase inhibitor BAY 43-9006. Oncogene 24:6861–6869

Yu JL, Coomber BL, Kerbel RS (2002) A paradigm for therapy-induced microenvironmental changes in solid tumors leading to drug resistance. Differentiation 70:599–609

Xian X, Hakansson J, Stahlberg A et al (2006) Pericytes limit tumor cell metastasis. J Clin Invest 116:642–651

Xin Y, Lyness G, Chen D et al (2005) Low dose suramin as a chemosensitizer of bladder cancer to mitomycin C. J Urol 174:322–327

Zeisberger SM, Odermatt B, Marty C et al (2006) Clodronate-liposome-mediated depletion of tumour-associated macrophages: a new high and highly effective antiangiogenic therapy approach. Br J Cancer 95:272–281

Zhang H, Vakil V, Braunstein M et al (2005) Circulating endothelial progenitor cells in multiple myeloma: implications and significance. Blood 105:3286–3294

Zhang Y, Song S, Yang F et al (2001) Nontoxic doses of suramin enhance activity of doxorubicin in prostate tumors. J Pharmacol Exp Ther 299:426–433

Zhang S, Yu H, Zhang L (2009) Clinical implications of increased lymph vessel density in the lymphatic metastasis of early-stage invasive cervical carcinoma: a clinical immunohistochemical method study. BMC Cancer 9:64

Zheng PP, Hop WC, Luider TM et al (2007) Increased levels of circulating endothelial progenitor cells and circulating endothelial nitric oxide synthase in patients with gliomas. Ann Neurol 62:40–48

Zimmermann K, Schmittel A, Steiner U et al (2009) Sunitinib treatment for patients with advanced clear-cell renal—cell carcinoma after progression on sorafenib. Oncology 76:350–354

Zorick TS, Mutsacchi Z, Bandos Y et al (2001) High serum endostatin levels in Down syndrome: implications for improved treatment and prevention of solid tumours. Eur J Hum Genet 9:811–814

Zwolak P, Jasinski P, Terai K et al (2008) Addition of receptor tyrosine kinase inhibitor to radiation increases tumor control in an orthotopic murine model of breast cancer metastasis in bone. Eur J Cancer 44:2506–2517

Index

D. Ribatti, *Morphofunctional Aspects of Tumor Microcirculation,*
DOI 10.1007/978-94-007-4936-8, © Springer Science+Business Media Dordrecht 2012

Printed by Printforce, the Netherlands